ARCHITECTS CONTRACTORS ENGINEERS

Guide To Construction COSTS

2006 Vol. XXXVII

Architects Contractors Engineers Guide to Construction Costs 2006, Volume 37.

ISBN 1-58855-069-9

ARCHITECTS CONTRACTORS & ENGINEERS
GUIDE TO CONSTRUCTION COSTS

EDITOR'S NOTE 2006

This annually published book is designed to give a uniform estimating and cost control system to the General Building Contractor. It contains a complete system to be used with or without computers. It also contains Quick Estimating sections for preliminary conceptual budget estimates by Architects, Engineers and Contractors. Square Foot Estimating is also included for preliminary estimates.

The Metropolitan Area concept is also used and gives the cost modifiers to use for the variations between Metropolitan Areas. This encompasses over 50% of the industry. This book is published annually to be historically accurate with the traditional May-July wage contract settlements and to be a true construction year estimating and cost guide.

The Rate of Inflation in the Construction Industry in 2004 was *8%*. Labor contributed a *4%* increase and materials rose *10%*.

The Wage Rate for Skilled Trades increased an average of *6%* in 2005. Wage rates will probably increase at a 6% average next year.

The Material Rate increased *10%* in 2005. The main increases were in asphalt, drywall, piping, steel and steel products, and insulation.

Construction Volume should stay level 2006. Housing will probably moderate, and Industrial and Commercial Construction should rise. Highway and Heavy Construction should stay flat.

The Construction Industry again has high inflation. Some Materials should inflate at a slower pace, and Labor should again inflate in 2006.

We are recommending using a *8%* increase in your estimates for work beyond July 1, 2005.

Don Roth
Editor

State	Metropolitan Area	Multiplier	State	Metropolitan Area	Multiplier
AK	ANCHORAGE	137	CA	REDDING	111
AL	ANNISTON	90		RIVERSIDE-SAN BERNARDINO	112
	AUBURN-OPELIKA	90		SACRAMENTO	114
	BIRMINGHAM	88		SALINAS	116
	DOTHAN	86		SAN DIEGO	113
	GADSDEN	86		SAN FRANCISCO	125
	HUNTSVILLE	88		SAN JOSE	122
	MOBILE	90		SAN LUIS OBISPO	110
	MONTGOMERY	86		SANTA CRUZ-WATSONVILLE	116
	TUSCALOOSA	88		SANTA ROSA	117
AR	FAYETTEVILLE-SPRINGDALE-ROGERS	80		STOCKTON-LODI	113
	FORT SMITH	86		VALLEJO-FAIRFIELD-NAPA	116
	JONESBORO	85		VENTURA	112
	LITTLE ROCK-NORTH LITTLE ROCK	87		SANTA BARBARA	115
	PINE BLUFF	87	CO	BOULDER-LONGMONT	96
	TEXARKANA	86		COLORADO SPRINGS	101
AZ	FLAGSTAFF	98		DENVER	101
	PHOENIX-MESA	98		FORT COLLINS-LOVELAND	94
	TUCSON	97		GRAND JUNCTION	96
	YUMA	99		GREELEY	94
CA	BAKERSFIELD	112		PUEBLO	98
	FRESNO	114	CT	BRIDGEPORT	112
	LOS ANGELES-LONG BEACH	115		DANBURY	112
	MODESTO	111		HARTFORD	111
	OAKLAND	120		NEW HAVEN-MERIDEN	112
	ORANGE COUNTY	114		NEW LONDON-NORWICH	109

State	Metropolitan Area	Multiplier	State	Metropolitan Area	Multiplier
CT	STAMFORD-NORWALK	115	KS	TOPEKA	93
	WATERBURY	111		WICHITA	93
DC	WASHINGTON	104	KY	LEXINGTON	93
DE	DOVER	104		LOUISVILLE	96
	WILMINGTON-NEWARK	105		OWENSBORO	94
FL	DAYTONA BEACH	92	LA	ALEXANDRIA	88
	FORT LAUDERDALE	96		BATON ROUGE	92
	FORT MYERS-CAPE CORAL	88		HOUMA	92
	FORT PIERCE-PORT ST. LUCIE	95		LAFAYETTE	90
	FORT WALTON BEACH	97		LAKE CHARLES	92
	GAINESVILLE	90		MONROE	88
	JACKSONVILLE	94		NEW ORLEANS	95
	LAKELAND-WINTER HAVEN	90		SHREVEPORT-BOSSIER CITY	89
	MELBOURNE-TITUSVILLE-PALM BAY	98	MA	BARNSTABLE-YARMOUTH	115
	MIAMI	96		BOSTON	118
	NAPLES	98		BROCKTON	114
	OCALA	95		FITCHBURG-LEOMINSTER	111
	ORLANDO	95		LAWRENCE	114
	PANAMA CITY	83		LOWELL	109
	PENSACOLA	88		NEW BEDFORD	114
	SARASOTA-BRADENTON	89		PITTSFIELD	110
	TALLAHASSEE	85		SPRINGFIELD	110
	TAMPA-ST. PETERSBURG-CLEARWATER	93		WORCESTER	109
	WEST PALM BEACH-BOCA RATON	96	MD	BALTIMORE	98
GA	ALBANY	85		CUMBERLAND	96
	ATHENS	87		HAGERSTOWN	91
	ATLANTA	95	ME	BANGOR	95
	AUGUSTA	83		LEWISTON-AUBURN	96
	COLUMBUS	84		PORTLAND	97
	MACON	87	MI	ANN ARBOR	106
	SAVANNAH	89		DETROIT	111
HI	HONOLULU	135		FLINT	105
IA	CEDAR RAPIDS	97		GRAND RAPIDS-MUSKEGON-HOLLAND	100
	DAVENPORT	100		JACKSON	104
	DES MOINES	101		KALAMAZOO-BATTLE CREEK	96
	DUBUQUE	95		LANSING-EAST LANSING	104
	IOWA CITY	100		SAGINAW-BAY CITY-MIDLAND	102
	SIOUX CITY	95	MN	DULUTH	106
	WATERLOO-CEDAR FALLS	94		MINNEAPOLIS-ST. PAUL	111
ID	BOISE CITY	99		ROCHESTER	106
	POCATELLO	97		ST. CLOUD	108
IL	BLOOMINGTON-NORMAL	106	MO	COLUMBIA	99
	CHAMPAIGN-URBANA	105		JOPLIN	95
	CHICAGO	115		KANSAS CITY	102
	DECATUR	104		SPRINGFIELD	97
	KANKAKEE	107		ST. JOSEPH	99
	PEORIA-PEKIN	106		ST. LOUIS	99
	ROCKFORD	106	MS	BILOXI-GULFPORT-PASCAGOULA	87
	SPRINGFIELD	104		JACKSON	86
IN	BLOOMINGTON	101	MT	BILLINGS	98
	EVANSVILLE	100		GREAT FALLS	99
	FORT WAYNE	100		MISSOULA	97
	GARY	108	NC	ASHEVILLE	82
	INDIANAPOLIS	104		CHARLOTTE	83
	KOKOMO	100		FAYETTEVILLE	84
	LAFAYETTE	100		GREENSBORO-WINSTON-SALEM-H POINT	83
	MUNCIE	100		GREENVILLE	83
	SOUTH BEND	101		HICKORY-MORGANTON-LENOIR	79
	TERRE HAUTE	100		RALEIGH-DURHAM-CHAPEL HILL	83
KS	KANSAS CITY	99		ROCKY MOUNT	79
	LAWRENCE	94		WILMINGTON	84

State	Metropolitan Area	Multiplier
ND	BISMARCK	95
	FARGO	96
	GRAND FORKS	94
NE	LINCOLN	92
	OMAHA	96
NH	MANCHESTER	100
	NASHUA	98
	PORTSMOUTH	92
NJ	ATLANTIC-CAPE MAY	110
	BERGEN-PASSAIC	113
	JERSEY CITY	112
	MIDDLESEX-SOMERSET-HUNTERDON	108
	MONMOUTH-OCEAN	111
	NEWARK	114
	TRENTON	111
	VINELAND-MILLVILLE-BRIDGETON	109
NM	ALBUQUERQUE	96
	LAS CRUCES	92
	SANTA FE	98
NV	LAS VEGAS	108
	RENO	107
NY	ALBANY-SCHENECTADY-TROY	102
	BINGHAMTON	100
	BUFFALO-NIAGARA FALLS	108
	ELMIRA	95
	GLENS FALLS	94
	JAMESTOWN	102
	NASSAU-SUFFOLK	115
	NEW YORK	132
	ROCHESTER	106
	SYRACUSE	103
	UTICA-ROME	103
OH	AKRON	103
	CANTON-MASSILLON	101
	CINCINNATI	98
	CLEVELAND-LORAIN-ELYRIA	106
	COLUMBUS	101
	DAYTON-SPRINGFIELD	99
	LIMA	101
	MANSFIELD	97
	STEUBENVILLE	104
	TOLEDO	103
	YOUNGSTOWN-WARREN	100
OK	ENID	88
	LAWTON	89
	OKLAHOMA CITY	91
	TULSA	90
OR	EUGENE-SPRINGFIELD	106
	MEDFORD-ASHLAND	104
	PORTLAND	108
	SALEM	107
PA	ALLENTOWN-BETHLEHEM-EASTON	105
	ALTOONA	103
	ERIE	103
	HARRISBURG-LEBANON-CARLISLE	101
	JOHNSTOWN	104
	LANCASTER	101
	PHILADELPHIA	114
	PITTSBURGH	104
	READING	103
	SCRANTON-WILKES-BARRE-HAZLETON	102

State	Metropolitan Area	Multiplier
PA	STATE COLLEGE	98
	WILLIAMSPORT	100
	YORK	102
RI	PROVIDENCE	109
SC	AIKEN	91
	CHARLESTON-NORTH CHARLESTON	85
	COLUMBIA	85
	FLORENCE	82
	GREENVILLE-SPARTANBURG-ANDERSON	84
	MYRTLE BEACH	90
SD	RAPID CITY	89
	SIOUX FALLS	90
TN	CHATTANOOGA	88
	JACKSON	87
	JOHNSON CITY	84
	KNOXVILLE	88
	MEMPHIS	93
	NASHVILLE	92
TX	ABILENE	86
	AMARILLO	87
	AUSTIN-SAN MARCOS	88
	BEAUMONT-PORT ARTHUR	88
	BROWNSVILLE-HARLINGEN-SAN BENITO	89
	BRYAN-COLLEGE STATION	87
	CORPUS CHRISTI	85
	DALLAS	91
	EL PASO	85
	FORT WORTH-ARLINGTON	91
	GALVESTON-TEXAS CITY	90
	HOUSTON	92
	LAREDO	78
	LONGVIEW-MARSHALL	83
	LUBBOCK	88
	MCALLEN-EDINBURG-MISSION	84
	ODESSA-MIDLAND	85
	SAN ANGELO	83
	SAN ANTONIO	89
	TEXARKANA	86
	TYLER	86
	VICTORIA	86
	WACO	86
	WICHITA FALLS	86
UT	PROVO-OREM	94
	SALT LAKE CITY-OGDEN	93
VA	CHARLOTTESVILLE	88
	LYNCHBURG	87
	NORFOLK-VA BEACH-NEWPORT NEWS	91
	RICHMOND-PETERSBURG	91
	ROANOKE	85
VT	BURLINGTON	98
WA	BELLINGHAM	112
	BREMERTON	110
	OLYMPIA	108
	RICHLAND-KENNEWICK-PASCO	106
	SEATTLE-BELLEVUE-EVERETT	111
	SPOKANE	108
	TACOMA	111
	YAKIMA	106
WI	APPLETON-OSHKOSH-NEENAH	101
	EAU CLAIRE	100
	GREEN BAY	101

State	Metropolitan Area	Multiplier
WI	JANESVILLE-BELOIT	101
	KENOSHA	103
	LA CROSSE	99
	MADISON	100
	MILWAUKEE-WAUKESHA	107
	RACINE	103
	WAUSAU	100

State	Metropolitan Area	Multiplier
WV	CHARLESTON	99
	HUNTINGTON	100
	PARKERSBURG	98
	WHEELING	100
WY	CASPER	92
	CHEYENNE	93

HOW TO USE THIS BOOK

Labor Unit Columns
> Units *include* all taxable (vacation) and untaxable (welfare and pension) fringe benefits.
> Units *do not include* taxes and insurance on labor (approximately 35% added to labor cost). (Workers Compensation, Unemployment Compensation, and Employers FICA.)
> Units *do not include* general conditions and equipment (approx. 10%).
> Units *do not include* contractors' overhead and profit (approx. 10%).
> Units are Union or Government Minimum Area wages.

Material Unit Columns
> Units *do not include* general conditions and equipment (approx. 10%).
> Units *do not include* sales or use taxes (approx. 5%) of material cost.
> Units *do not include* contractors' overhead and profit (approx. 10%).
> Units are mean prices FOB job site.

Cost Unit Columns (Subcontractors)
> Cost is complete with labor, material, taxes, insurance and fees of a subcontractor. This also applies to Quick Estimating sections.
> Units *do not include* general contractors' overhead or profit.

Quick Estimating Sections - For Preliminary and Conceptual Estimating
> Includes all labor, material, general conditions, equipment, taxes, insurance and fees of contractors combined for items not covered in Cost Unit columns.

Construction Modifiers - For Metro Area Cost Variation Adjustments
> Labor items are percentage increases or decreases from 100% for labor units used in the book's labor unit columns. (See first page of each Division.)
> Material and Labor combined are percentage increases and decreases from 100% for metro variations used in the book's Unit Cost columns, Quick Estimating sections and Sq.Ft. Cost Section - see Page 2.
> Recommend Unit Costs of 10% less than Book Units for Residential Construction.

EXAMPLE:	UNIT	LABOR	MATERIAL	COST
2" x 4" Wood Studs & Plates - 8' @ 16" O.C.	BdFt	.67	.55	1.22
Add 35% Taxes & Insurance on Labor		.23	-	.23
Add 5% Material Sales Tax		-	.03	.03
				1.48
Add 10% General Conditions & Equipment				.15
				1.63
Add 10% Overhead & Profit				.17
BdFt Cost - Bid Type Estimating			BdFt	1.80
or SqFt Wall Cost with 3 Plates - Quick Estimating (page 6A-14)			or SqFt	1.40

Purchasing Practice (usual) in the industry for each section of work is indicated immediately after section heading in parenthesis and defined as follows:

L&M - Labor and Material Subcontractor - Firm bid to G.C. according to plans and specifications for all labor and material in that section or subsection.

M - Material Subcontractor - Firm bid to G.C. according to plans and specifications for all material in that section or subsection.

L - Labor Subcontractor - Firm bid to G.C. to *install only* material furnished according to plans and specifications by others.

S - Material Supplier - Firm Price for materials delivered only.

E - Equipment Supplier - Firm Price for Equipment Rented or Purchased.

Installation Practice by particular trades are listed behind each section heading and are according to "Agreement and Decisions Rendered Affecting the Building Industry," by the A.F.L. Building and Construction Trades Department, A.F.L.-C.I.O. Area variations, jurisdictional problems and multiple jurisdictions are possible.

General Requirements amount to approximately 6% to 15% of total job costs for General Construction and 5% to 10% for Residential Construction. All items below are included in G.R. except 0101.32, .33, and .34 (taxes and insurance on labor). See Square Foot Costs (SF Section) for more accurate Dollar Value of Job Costs.

0101.0 GENERAL CONDITIONS - Specified (AIA Doc A201)

			UNIT	LABOR	MATERIAL
.1	SUPERINTENDENT Job Size/Complexity		Week	$1,900	-
.2	PERMITS AND FEES				
	Concrete and Commercial	- Approx. Average	MCuFt	-	$3.70
	Residential (Finish Area)	- Approx. Average	SqFt	-	$ 1.10
	Remodeling	- Approx. Average	M$	-	$6.50
	Varies with Each Permit Division				
.3	TAXES (Art. 3.6)				
.31	Sales and Use Tax (Varies as to State)		%		Avg. 5.7%

		Payroll Max	Per $100 or % Gross Payroll	
.32	Unemployment Compensation	to $ 22,000	State Maximum	9.1%
	(Varies as to State)		State Minimum	.7%
.33	Unemployment Compensation	to $ 7,000	Fed. Maximum	.8%
.34	FICA - Employer's Share	to $ 87,900		6.2%
	Medicare	No Limit		1.45%

.4	INSURANCE - CONTRACTOR'S LIABILITY			
	(Worker's Compensation, P.L. & P.D.)			
	Classification Avg - Contractors Experience		Gross Payroll	14.0%
	Clerical - Office	(#8810)	" "	1.0%
	Supervision	(#5606)	" "	2.5%
	Excavation	(#6217)	" "	12.0%
	Concrete - Structural - N.O.C.	(#5213)	" "	20.0%
	Slabs on Ground, Flat	(#5221)	" "	8.0%
	Masonry	(#5022)	" "	12.0%
	Iron and Steel - N.O.C.	(#5057)	" "	32.0%
	Carpentry - N.O.C. Commercial	(#5403)	" "	19.0%
	Interior & Dwellings	(#5437)	" "	12.0%
	Wallboard	(#5445)	" "	13.0%
	Plumbing			7.0%
	Electrical			6.0%
	Wrecking (Demolition)			40.0%
	PL. and PD.			1.5%
	Add for Assigned Risk			75%

			UNIT	LABOR	MATERIAL
.5	INSURANCE - HOLD HARMLESS				
	(By Owner) (Art. 11.2) (Contract Amount)		M$		$ .80
.6	INSURANCE - ALL RISK (Fire/Ext.Cov./Theft/etc.)				
	(By Owner) (Art. 11.2) ($1,000 Deductible)				
	Fire Proof - All Concrete		Annual)		$2.20
	Fire Resistive - Covered Steel		Insurable)		$2.75
	Ordinary - Masonry, Ext. Walls & Joists		Value)		$3.80
	Frame, Wood		per M$)		$6.50
	Deduct for $2,500 Deductible				10%
	Add for Light Protection Areas - Approx.				50%
.7	CLEANUP (Art. 3.15)				
.71	Progress		Bldg - SqFt	$.18	$ .05
.72	Glass, 2 Sides		Glass Area - SqFt	$.35	-
.8	PERFORMANCE & PAYMENT BOND		(If Specified) (Art. 11.4)		
	$ -0- to $ 99,999		Job Bureau Comp. Contr. Amount		2.50%
	$ 100,000 to $ 499,999				1.50%
	$ 500,000 to $2,499,999				1.00%
	$2,500,000 to $4,999,999				.75%
	$5,000,000 to $7,499,999				.70%
	$7,500,000 Up				.65%
.9	SAFETY OF PERSONS & PROPERTY - Job Condition				
.10	TESTS AND SAMPLES - See Appropriate Division				

0102.0 SPECIAL CONDITIONS (Changes to G.C. 0101 and Temporary Utilities & Facilities, usually Specified)

			UNIT	LABOR	MATERIAL
.1	TEMPORARY OFFICE				
	Fixed - 10' x 16" Wood		Each	1,500	2,700
	Mobile (Special Construction) 8' x 32'		Each	-	4,300
			Month	-	210
	Mobile (Special Construction) 10' x 50'		Each	1,200	7,500
			Month	-	310
	Mobile (Special Construction) 12' x 60'		Each	1,400	10,800
			Month	-	420
	Add for Delivery & Pickup		Each	-	420
.2	TEMPORARY STORAGE SHED -12' x 24'		Each	1,900	4,000
.3	TEMPORARY SANITARY FACILITIES - Chemical Toilet		Month	-	105
.4	TEMPORARY TELEPHONE (No Long Distance) - Per line		Month	-	100
	TEMPORARY TELEPHONE		Initial	-	320
	TEMPORARY TELEPHONE - Cellular		Month	-	100
	TEMPORARY FAX		Initial	-	1,200
.5	TEMPORARY POWER AND LIGHT				
	Initial Service 110V		Job	-	1,400
	220V		Job	-	2,000
	440V		Job	-	4,000
	Monthly Service and Material		Month	-	260
	Temporary Wiring and Lighting-Total Job		SqFt	-	.09
.6	TEMPORARY WATER -Initial Service		Initial	Job Condition	
	Monthly Service		Month	-	100.00
.7	TEMPORARY HEAT (after Enclosed)				
	See Divisions 3 & 4 for Temporary Heat, Concrete & Masonry				
	See Division 1A for Equipment Rental Rates				
	Heaters:				
	Steam & Unit - Heater Rent - SqFt Htd Area		Month	-	.12
	Piping - SqFt Htd Area		Month	-	.14
	Condensate - SqFt Htd Area		Month	-	.19
	Mobile Home Heater Rent - SqFt Htd Area		Month	-	.12
	Fuel SqFt Htd Area		Month	-	.25
	Tending 8 - Hour/Day		8 Hrs	-	250.00
.8	TEMPORARY ENCLOSURES AND COVERS				
	Per Door - Wood		Each	34.00	46.00
	Per Window - Polyethylene and Frame		Each	15.00	12.50
	Walls - Frame and Reinforced Poly - 1 use		SqFt	.23	.19
	2 uses		SqFt	.21	.15
	Frame and 1/2" Plywood - 1 use		SqFt	.64	1.00
	2 uses		SqFt	.53	.65
	Floors- 2" Fiberglass & 7 mil Poly- 1 use		SqFt	.13	.47
.9	TEMPORARY ROADS & RAMPS - No Surfacing		SqYd	Job Condition	
.10	TEMPORARY DEWATERING - No Tending				
	Pumping - 2" Pump		Day	-	95.00
	3" Pump		Day	-	105.00
	4" Pump		Day	-	125.00
	Add for Tending				
	Well Points - LnFt Header		Month	-	105.00
	Add Each Month		Month	-	80.00
.11	TEMPORARY SIGNS - 4' x 8'		Each	100.00	450.00
.12	TEMP. FENCING, BARRICADES & WARNING LIGHTS				
	6' Chain Link - Rented (Erected and Removed)		LnFt	1.25	5.50
	Add for Barbed Wire		LnFt	.35	1.00
	Add for Gates		LnFt	8.00	20.00
	Snow Fence - New		LnFt	.40	1.25
	Concrete Barriers (Rented - Month)		LnFt	-	4.00
	Flasher Warning Light (Rented - Month)		Each	-	20.00
	8' Barricade (Rented - Month)		Each	-	45.00
	12' Barricade (Rented - Month)		Each	-	55.00

0103.0	MISCELLANEOUS CONDITIONS (Non-Specified)	UNIT	LABOR*	MATERIAL
.1	FIELD ENGINEERING AND LAYOUT	Week		Job Cond.
.2	TIMEKEEPERS AND ACCOUNTANTS	Week		Job Cond.
.3	SECURITY	Week		Job Cond.
.4	OPERATING ENGINEERS FOR HOISTING	Week		Job Cond.
.5	TRAVEL AND SUBSISTENCE - Field Employees			Job Cond.
.6	TRAVEL AND SUBSISTENCE - Office Employees			Job Cond.
.7	OFFICE SUPPLIES AND EQUIPMENT	Week		25.00
.8	COMMUNICATIONS EQUIPMENT			
	Radio-Multi Channel Hand Unit with Charger & Battery	Each	-	1,350
	Base Unit - F.M. with Antenna and Mike	Each	-	1,500
.9	HOISTING EQUIPMENT (Towers, Cranes and Lifts)			
.91	Towers and Hoists - Portable Self-Erecting Style			
	Skip (Based on 100' Unit) -3000# to 5000# Capacity	Month	-	2,300
	2000# to 3000# Capacity	Month	-	2,100
	Add to Heights Above 100'	LnFtMo		21
	Towers - Fixed (based on 100') - 30' Override			
	Tubular Frame - Double Well - 5000# Capacity	Month	-	2,000
	Installation - Double Well - Subcont - Up	LnFt	65	-
	Double Well - Down	LnFt	50	-
	Add for Concrete Bucket and Hopper	Month	-	400
	Add for Chicago Boom	Month	-	230
	Add for Gates, Signals, etc (Safety)	Job	750	1,200
	Hoist Engine, Gas Powered - 2 Drum	Month	-	1,800
	Towers and Hoists - Personnel 6000# Capacity			
	Based on 100' Unit - 8 1/2' Override	Month	-	3,500
	Add Above 100'	LnFt	-	35
	Installation - Up	LnFt	-	40
	Down	LnFt	-	38
.92	Cranes - Climbing - 20' Override			
	Based on 125' Bldg-130' Boom, 4000#	Month	-	6,500
	to 200' Boom, 7000#	Month	-	9,000
	230' Boom, 7000#	Month	-	10,500
	Add for LnFt Tower above 125'	Month	-	40
	Installation			
	Foundation	Each	-	12,000
	440-Volt Transformer Install	Each	-	2,500
	Installation - Up & Down	LnFt	-	110
	Up	Each	-	15,000
	Down	Each	-	15,000
	Avg - Up, Down & Haul - 125' Tower	Each	-	30,000
	Add per Lift - Jacking	LnFt	-	50
	Add per Floor Pattern	Each	-	1,000

.93	Cranes - Mobile Lifting (no Operator)	HOUR	DAY	MONTH
	15-Ton Hydraulic	110	600	3,800
	20-Ton	120	650	4,000
	25-Ton	125	700	6,600
	30-Ton	140	750	7,000
	45-Ton	160	800	7,800
	75-Ton Cable	175	900	8,700
	90-Ton	200	1,100	9,500
	125-Ton	240	1,500	13,000

.94	Fork Lifts - Construction	UNIT	LABOR	MATERIAL
	Four-Wheel Drive - 28' 5000# Capacity	Month		2,300
	36' 6000# Capacity	Month		2,400
	37' 6000# Capacity	Month		2,500
	42' 6000# Capacity	Month		2,600
	42' 8000# Capacity	Month		2,700

Add for Local Travel & Setup Time - 4 Hours

*Subcontract

0103.0   MISCELLANEOUS CONDITIONS, Cont'd...

.95   Platform Lifts

			UNIT	LABOR	MATERIAL
Scissor Lifts – Electric		26'	Month		615
		32'	Month		800
Telescoping Boom Lifts		46'	Month		4,800
		66'	Month		2,450
Articulating Boom Lifts		86'	Month		5,100
		36'	Month		1,500
		51'	Month		1,900
		66'	Month		2,450
.10	TRUCKS AND OTHER JOB VEHICLES (Site Use)				
	$ 500,000 Job		JobAvg		1,700
	$1,000,000 Job		JobAvg		6,200
	$5,000,000 Job		JobAvg		14,500
	1-Ton Pickup		Month		625
	2-Ton Flatbed		Month		880
	2 1/2-Ton Dump		Month		1,500
.11	MAJOR EQUIPMENT (See Div. 2,3,4,5,& 6 if Subbed)				
	(See 1-6A & 7A: Rental Rates, Purchase Prices)				
	$ 100,000 Job		JobAvg		4,000
	$ 500,000 Job		JobAvg		7,800
	$1,000,000 Job		JobAvg		14,500
	$5,000,000 Job		JobAvg		58,000
	As a % of Labor (No Hoisting Equipment)				5%
.12	SMALL TOOLS AND SUPPLIES				
	(See 1-6A & 7A: Rental Rates, Purchase Prices)				
	$ 100,000 Job		JobAvg		1,400
	$ 500,000 Job		JobAvg		3,800
	$1,000,000 Job		JobAvg		6,500
	$5,000,000 Job		JobAvg		30,000
	As a Percentage of Labor		JobAvg		2%
.13	HAULING EQUIPMENT AND TOOLS				
	$ 100,000 Job		JobAvg	950	600
	$ 500,000 Job		JobAvg	1,800	1,500
	$1,000,000 Job		JobAvg	3,500	3,000
	$5,000,000 Job		JobAvg	11,000	8,000
.14	GAS AND OIL FOR EQUIPMENT				
	$ 500,000 Job		JobAvg		2,000
	$1,000,000 Job		JobAvg		2,600
	$5,000,000 Job		JobAvg		11,000
	As a Average of Equipment Cost		JobAvg		17%
.15	REPAIRS TO EQUIPMENT AND TOOLS				
	$ 100,000 Job		JobAvg		600
	$ 500,000 Job		JobAvg		1,200
	$1,000,000 Job		JobAvg		2,200
	$5,000,000 Job		JobAvg		7,200
	As an Average of Equipment Cost				10%
.16	PHOTOGRAPHS - 2 Views, 4 Copies		Month		140
.17	SCHEDULING - C.P.M., Pert, Network, Analysis				
	$ 500,000 Job		JobAvg		4,500
	$1,000,000 Job		JobAvg		8,000
	$5,000,000 Job		JobAvg		14,000
	Add for Update, Monitor		Each		750
.18	COMPUTER COSTS		Employee/Wk		10
.19	PERFORMANCE BOND - SUBCONTRACTORS		ContrAmt		1.5%
.20	PUNCH LIST AND FINAL OUT		ContrAmt		.5%

TOOLS & EQUIPMENT

See 103.9 for Towers, Cranes and Lift Equipment	Approximate Cost New	Fair Rental* Rate per Mo.
Div. 1 General Conditions		
Engineering Instruments - Transits	1,229	250
Levels 18"	998	230
Fans - 15,000 CFM - 36"	1,155	400
Generators - 3500 Watt - Portable	1,260	375
5000 Watt - Portable	1,628	425
Heaters, Temp- Oil 100M to 300 BTU	1,365	450
LP Gas 90M BTU	378	200
150M BTU	420	240
350M BTU	746	350
Laser - Rotating	4,800	600
Light Stands - Portable - 500 watt Quartz	704	260
Pumps- Submersible 1/2 HP Elect 2"	441	320
Centrifugal 2" 8,000 Gal	1,019	490
3" 17,000 Gal	1,050	550
Diaphragm 2" - 3" x 3" - Pneumatic	1,470	460
3" - 3" x 3" - Gas	1,523	590
Submersible - 5 HP 3"	1,575	500
Hose - Discharge - 3" - 25'	105	55
Suction - 3" - 20'	137	70
Vacuum Cleaners - Wet Dry - 5 Gal	630	160
Div. 2 Excavation (Including Demolition)		
Air Tools- Breakers Light Duty - 35 to 40#	1,008	260
Heavy Duty - 60 to 90#	1,386	340
Rock Drill - 40#	1,869	350
Chipping Hammer - 7# to 12#	809	275
Clay Digger	756	250
Hose - 3/4" - 50'	137	60
Compressors 125 cfm Gas - Tow	11,235	800
185 cfm Diesel - Trailer Mounted	14,280	825
26' Conveyor - 2 HP Electric	9,608	980
Electric Breakers - 60#	2,205	410
Jacks - Hydraulic - 20 Ton - 11 HP	840	320
Screw - 20 Ton	378	85
Saws - Chain 18"	578	395
Concrete Asphalt 14" Blade	1,680	250
24" Plate - Vibratory Gas 170#	2,205	645
Tampers - Rammer Gas 135#	2,415	740
Div. 3 Concrete		
Bucket & Boot- 1 CuYd	2,153	480
1 1/2 CuYd	2,730	490
2 CuYd	3,570	750
Buggies - Manual - 6 to 8 CuFt	494	180
Power Operated - 16 CuFt	6,300	850
Conveyors - 32' Hydraulic and Belt - Gas	9,870	1,220
50' Hydraulic and Belt - Gas	21,840	2,625
Core Drill - Diamond - 15 Amp	3,413	550
with Trailer	9,870	900
Crack Chaser	4,410	1,350
Drill - Cordless 3/8" Reversing	305	100
Heavy Duty 1/2" Reversing	389	150
3/4" Reversing	704	195
Sander- Hand - 9"	347	150
Grinder- Ceiling 7' to 11' - Elec	2,310	440
Floor - Elec - 1 1/2 HP	3,203	650
Hammer - Drill - 3/4" Variable Speed	494	190
Roto Chipping - 1 1/2"	525	490
Hopper - Floor 30 CuFt	1,155	230
Mixer - 6 CuFt	2,415	480
Planer - 10'	4,515	720
Scarifier & Scabblers - Single Bit	1,155	725
5 Bit Floor - 175 cfm	8,190	2,200
Stripper - Floor	3,938	430

*Weekly Rent 40% of monthly, Daily Rent 15% of monthly, Hourly Rent 2% of monthly.

TOOLS & EQUIPMENT

		Approximate Cost New	Fair Rental* Rate per Mo.
Div. 3	**Concrete, Cont'd...**		
	Saws - Rotary Hand - 7 1/4"	190	130
	9 1/4" - 8 1/4"	235	140
	Table 14" Radial Arm 3 HP	1,800	300
	Floor - Gas 11 HP	1,550	650
	Gas 18 HP	4,300	690
	Electric 8 HP 14"	1,820	650
	Cut off 14" Electric	710	475
	Screeds - Vibrating 6'	900	675
	12 1/2' Beam Type	1,950	685
	25' Beam Type	4,200	780
	Troweling Machines - Gas 36" 4 Blade 7 HP	1,700	500
	48" 4 Blade 6 HP	1,950	560
	Trunks & Tremmies - 12" and Hopper	180	65
	Vibrators - Flexible Head - 1 HP	1,100	330
	2 HP	1,200	360
	High cycle - 180 Cycle	3,200	550
	Back Panel 2 1/2 HP	750	675
	Wheelbarrows	115	60
Div. 4	**Masonry**		
	Hoist Electric - One ton	1,050	170
	Mixer - Mortar - 6 Cu Ft - Gas 7 HP	2,000	430
	Elec 1 1/2 HP	2,120	400
	8 Cu Ft - Gas 7 HP	2,400	450
	12 Cu Ft - Gas or Electric	3,600	940
	Pallet Jack	740	300
	Pump Grout - Hydraulic 10 GPM	6,200	1,400
	Saws - Table - 1 1/2 HP - 14"	2,150	400
	5 HP - 14"	3,100	450
	Scaffold - 5' - 6" with Braces	75	3.60
	6' - 6" with Braces	80	4.00
	Wheels	85	11.00
	Brackets	22	2.00
	Plank - Aluminum	90	5.00
	16' Laminated	35	3.00
	Base - Adj	25	1.60
	Splitter - 10,000#	1,700	380
	Rolling Towers	350	100
	Swing Stage - Motorized with deck-cables-safety	9,000	1,200
Div. 6	**Carpentry**		
	Door Hanging Eq. (Router, Mortiser, Template)	300	100
	Drills - Heavy Duty (See Div. 3)		
	Fasteners - Power - Drywall	150	45
	Hammer - Drill 1/2"	310	170
	Miter Box - 14" Power	410	330
	Nailers and Staplers - Spot with compressor	1,800	280
	Planers	500	150
	Sanders - Disc 9"	260	100
	Belt 3" - Heavy Duty	300	125
	Floor	420	330
	Saws - Band	300	250
	Chain	350	330
	Rotary, Hand - 7 1/4"	190	95
	8 1/4"	220	130
	Table - Fixed 10"	400	140
	Jig	200	80
	Reciprocating	450	220
	Cut-off - 12" Elec.	2,000	375
	Scaffold - Rolling	450	110
	Screwdriver - Variable Speed - Drywall	250	110
	Stud Drivers - Low Velocity	280	120
	High Velocity	400	140
	Wrench - 1/2" Impact - Air	290	140
	1" Impact - Air	600	485

*Weekly Rent 40% of monthly, Daily Rent 15% of monthly, Hourly Rent 2% of monthly.

0201.0 DEMOLITION & CLEARING (Op.Eng., Truck Dr. & Lab.)

		UNIT	COST with MACHINE
.1	STRUCTURE MOVING - Wood Frame	SqFt	15.00
	With Masonry	SqFt	16.00
.2	CLEARING AND GRUBBING		
	Trees - Trunk	Dia/Inch	15.00
	Stump - Removed	Dia/Inch	12.00
	Chipped - Below Grade	Dia/Inch	7.00
	Add for Hauling Away and Dumping	Dia/Inch	6.00
	Shrubs & Brush - Light	Acre	2,300.00
	Medium	Acre	3,200.00
	Heavy	Acre	4,300.00
.3	DEMOLITION AND LOAD (See .34 for Disposal)		
.31	Total Building		
	Wood - Mainly Housing	CuFt/Bldg	.32
	Masonry - Load Bearing	CuFt/Bldg	.37
	Concrete - Frame	CuFt/Bldg	.47
	Steel - Frame	CuFt/Bldg	.37
	Add for Floor Above 2 Stories	CuFt/Bldg	.07

.32 Selective Building Removals (No Cutting/Disposal)
   (See .35 - Cut and Drill)
   (See .34 - Disposal/Haul Away)
   (See .33 - Site Removal)

	Unit	Labor Only	Machine Only	Unit	Labor Only	Machine Only
Concrete:						
8" Walls Reinforced	CuYd	270.00	100.00	SqFt	6.90	1.95
Non-Reinforced	CuYd	220.00	87.00	SqFt	5.25	1.70
12" Walls Reinforced	CuYd	370.00	77.00	SqFt	13.40	3.00
Footings - 24" x 12"	CuYd	300.00	170.00	LnFt	11.00	6.30
6" Struc. Slab Reinforced	CuYd	235.00	70.00	SqFt	4.30	1.20
8" Struc. Slab Reinforced	CuYd	250.00	90.00	SqFt	6.10	1.80
4" Slab on Ground Reinforced	CuYd	450.00	56.00	SqFt	2.00	.66
Non-Reinforced	CuYd	400.00	50.00	SqFt	1.55	.63
6" Slab on Ground Reinforced	CuYd	280.00	47.00	SqFt	2.50	.84
Non-Reinforced	CuYd	240.00	44.00	SqFt	1.75	.80
9" Stairs Reinforced	CuYd	320.00	99.00	SqFt	7.25	2.75
Masonry:						
4" Brick or Stone Walls	CuYd	115.00	34.00	SqFt	1.40	.42
and 8" Backup	CuYd	210.00	71.00	SqFt	2.40	.73
Block or Tile Partitions	CuYd	89.00	32.00	SqFt	1.10	.38
6" Block or Tile Partitions	CuYd	72.00	24.00	SqFt	1.25	.46
8" Block or Tile Partitions	CuYd	63.00	44.00	SqFt	1.50	.56
12" Block or Tile Partitions	CuYd	167.00	60.00	SqFt	1.90	.68
Add for Plastered Type				SqFt	.42	.13

Misc - Hand Work (Add for Equipment & Scaffold)	UNIT	Labor	
Acoustical Ceilings - Attached (Incl. Iron)	SqFt	.65	-
Suspended (Incl. Grid)	SqFt	.40	-
Asbestos - Ceilings and Walls	SqFt	16.00	-
Columns and Beams	SqFt	40.00	-
Pipe	LnFt	50.00	-
Tile Flooring	SqFt	1.85	-
Cabinets and Tops	LnFt	11.00	-
Carpet	SqFt	.34	-
Ceramic and Quarry Tile	SqFt	1.30	-
Doors and Frames - Metal	Each	57.00	-
Wood	Each	51.25	-
Drywall Ceilings - Attached (Including Grid)	SqFt	.80	-
Wood or Metal Studs - 2 Sides	SqFt	.90	-
Paint Removal - Doors and Windows	SqFt	.80	-
Walls	SqFt	.64	-
Plaster on Wood or Metal Studs	SqFt	1.20	-
Ceilings - Attached (Incl. Iron)	SqFt	1.15	-
Roofing - Built-Up	SqFt	1.10	-
Shingles - Asphalt and Wood	SqFt	.36	-
Terrazzo Flooring	SqFt	1.85	-
Vinyl Composition Flooring (No Mastic Removal)	SqFt	.38	-
Wall Coverings	SqFt	.70	-
Windows - Metal or Wood	Each	36.50	-
Wood Flooring	SqFt	.39	-

0201.0   DEMOLITION & CLEARING, Cont'd...

.33   Site Removals
(See .34 for Disposal)
(See .35 for Cutting Additions)

	UNIT	(1) LABOR ONLY	MACHINE ONLY
Concrete:4" Sidewalks - Reinforced	SqFt	2.10	.68
Non -Reinforced	SqFt	1.65	.65
6" Drives - Reinforced	SqFt	2.45	.86
Non-Reinforced	SqFt	1.90	.90
Curb - 6" x 18"	LnFt	5.70	1.65
and Gutter	LnFt	9.25	2.20
Asphalt:2"	SqFt	1.30	.42
Curb	LnFt	3.00	.73
Fencing: 8" Metal	LnFt	1.90	.46

.34   Disposal or Haul Away (Including Truck or Box & Driver)

	UNIT		COST
Haul - No Dump Charges Included			
Concrete Truck Measure - 75¢ a minute or	CuYd/Mile		.83
Masonry - 75¢ a minute or	CuYd/Mile		.80
Wood - 75¢ a minute or	CuYd/Mile		.75
Box Rent (Included Above) - 30 Cu.Yd.	Box		390.00
Clean Material - 20 Cu.Yd.	Box		360.00
10 Cu.Yd.	Box		310.00
Dump Charges - Concrete	CuYd		7.40
Building Materials	CuYd		10.80
Tree and Brush Removal	CuYd		15.25

.35   Saw Cutting/Core Drilling/Torch Cutting/Blasting

	UNIT		COST
Saw Cutting (includes Operator):			
Slabs - Asphalt	LnFt	per inch	1.25
Concrete - Structural	LnFt	per inch	2.30
Hollow	LnFt	per inch	.95
Slabs on Grd/Non - Rein.	LnFt	per inch	.83
/with Mesh	LnFt	per inch	1.05
Walls - Concrete - Reinforced	LnFt	per inch	3.50
Non - Reinforced	LnFt	per inch	2.80
Masonry - Hollow	LnFt	per inch	2.00
Solid	LnFt	per inch	2.65
Core Drilling:			
Solid Slab - 2" Diameter	Inch		4.80
3" Diameter	Inch		5.50
4" Diameter	Inch		7.25
6" Diameter	Inch		10.00
8" Diameter	Inch		13.50
Hollow Core Slab - 2" Diameter	Inch		3.00
3" Diameter	Inch		3.15
4" Diameter	Inch		4.50
6" Diameter	Inch		8.40
Torch Cutting - Steel per Inch	LnFt		5.80
Blasting - Unrestricted	CuYd		120.00
Restricted	CuYd		215.00
Cut Openings:			
Concrete (see .32)	SqFt		-
Sheet Rock Partitions	SqFt		3.15
Wood Floors	SqFt		2.60
No Disposal or Haul Away Included (see .34)			

(1) With Pneumatic Tools

0202.0 EARTHWORK (Op. Eng., Truck Driver & Lab.)

.1 GRADING

Machine

	UNIT	COST(1)	UNIT	COST(1)
Strip Top Soil - 4"	CuYd	5.50	SqYd	.62
Spread Top Soil				
4" Site Borrow	CuYd	6.10	SqYd	.69
Off-Site Borrow ($13 CuYd)	CuYd	23.70	SqYd	2.60

Rough Grading - Cut and Fill	UNIT	SOFT(1)	MED.(1)	HARD(1)
Dozer - To 200'	CuYd	4.00	4.30	4.80
To 500'	CuYd	4.50	5.20	8.00
Scraper - Self Propelled - To 500'	CuYd	3.70	4.00	4.60
To 1,000'	CuYd	4.50	5.20	6.20
Grader - To 200'	CuYd	4.00	4.50	6.50
Add for Compaction (See 0202.4)				
Add for Hauling Away (See 0202.5)				
Add for Truck Haul on Site	CuYd	3.10	-	-
Add for Borrow Brought In				
Loose Fill – 3.00 Ton	CuYd	11.10	-	-
Crushed Stone - 5.50 Ton	CuYd	14.60	-	-
Gravel - 4.30 Ton	CuYd	12.90	-	-

Hand and Machine

	UNIT	HAND LABOR	MACHINE
Fine Grading - 4" Fill - Site	CuYd	13.20	7.00
	or SqFt	.16	.09
4" Fill - Building	CuYd	24.00	11.50
	or SqFt	.30	.14

.2 EXCAVATION - Dig & Cast or Load (No Hauling)

Soil

	UNIT		
Open - Soft (Sand) with Backhoe or Shovel	CuYd	24.00	3.70
Medium (Clay)	CuYd	33.00	4.55
Hard	CuYd	49.00	6.50
Add for Clamshell or Dragline	CuYd	-	1.50
Deduct for Front End Loader	CuYd	-	1.25
Deduct for Dozer	CuYd	-	.30
Trench or Pocket - Soft with Backhoe or Shovel	CuYd	25.00	5.50
Medium	CuYd	33.00	7.25
Hard	CuYd	47.00	8.60
Add for Clamshell or Dragline	CuYd	-	1.40
Augered - 12" diameter	CuYd	54.00	10.30
24" diameter	CuYd	41.00	7.00
Add for Frost	CuYd	-	8.20
Rock - Soft - Dozer	CuYd	-	36.00
Hammer	CuYd	-	90.00
Medium - Blast	CuYd	-	110.00
Hammer	CuYd	-	180.00
Hard - Blast	CuYd	-	240.00
Hammer	CuYd	-	330.00
Swamp	CuYd	-	10.30
Underwater	CuYd	-	13.50

.3 BACKFILL (Not Compacted and Site Borrow)	CuYd	11.50	3.00
Add for Off-Site Borrow	CuYd	-	11.00

Add to <u>All Above</u> for Mobilization Average   5%

---

(1) Machine and Operators

0202.0 EARTHWORK, Cont'd...

		UNIT	LABOR [1]	COST [2]
.4	COMPACTION (Incl. Backfilling & Site Borrow)			
	Labor Only			
	Building to 85 - 24" Lifts	CuYd	16.50	-
	to 95 - 12" Lifts	CuYd	19.20	-
	to 100 - 6" Lifts	CuYd	23.00	-
	Labor Compaction & Machine Backfill Combined			
	Building to 85 - 24" Lifts	CuYd	10.00	3.20
	to 90 - 18" Lifts	CuYd	10.50	3.50
	to 95 - 12" Lifts	CuYd	11.20	3.40
	to 100 - 6" Lifts	CuYd	12.30	3.80
	Subgrade (Including Grading and Rollers)	CuYd	-	4.35
	Machine Only (No Grading)			
	Site to 85 - 24" Lifts	CuYd		3.60
	to 90 - 18" Lifts	CuYd		3.70
	to 95 - 12" Lifts	CuYd		4.35
	to 100 - 6" Lifts	CuYd		5.50
	Add to Above for Off-site Borrow	CuYd		12.00
.5	DISPOSAL (Haul Away) Country	CuYd/Mile		.76
	City	CuYd/Mile		.88
.6	TESTS			
	Field Density - Compaction	Each		40.00
	Soil Gradation	Each		59.00
	Density (Proctor)	Each		96.00
	Moisture and Specific Gravities	Each		105.00
	Hydrometer (Sand, Silt and Clay)	Each		11500
	Rock -2" Core	LnFt		72.00
	4" Core	LnFt		120.00
	Add for Soil Penetration	LnFt		8.20
	Add for Mileage	Mile		.54

0203.0 PILE AND CAISSON FOUNDATIONS (L&M)
(Operating Engineers and Iron Workers)

		UNIT	COST
.1	CONCRETE		
.11	Augured or Cast in Place -12"	LnFt	23.50
.12	Precast Prestressed -10"	LnFt	20.80
	12"	LnFt	24.00
	14"	LnFt	26.00
	16"	LnFt	32.00
.13	Pipe Casing Pulled - Cast - in - Place - 10"	LnFt	22.00
	12"	LnFt	23.00
.2	STEEL		
.21	Steel Pipe -10"	LnFt	27.00
	12"	LnFt	31.00
	14"	LnFt	36.00
	16"	LnFt	39.00
	Concrete - Filled - 10"	LnFt	30.00
	12"	LnFt	37.00
	14"	LnFt	44.00
	16"	LnFt	50.00
.22	Steel H Section 8" - 36#	LnFt	25.00
	10" - 42#	LnFt	28.00
	12" - 53#	LnFt	34.00
	14" - 73#	LnFt	36.00
	14" - 89#	LnFt	45.00
.23	Helical Pier Foundation		
	8" -10" - 12"	LnFt	26.00
	10" - 12" - 14"	LnFt	28.00
.3	WOOD (Treated) to 40' (Incl. Tests & Cutoffs)		
	10"	LnFt	19.00
	12"	LnFt	20.50
	14"	LnFt	23.00
	16"	LnFt	24.00
	Add for Mobilization		10%

[1] Hand Work
[2] Machine and Operators

		UNIT	COST
0203.0	**PILE & CAISSON, Cont'd...**		
.4	CAISSONS -20" Concrete Filled	LnFt	35.00
	24"	LnFt	41.00
	30"	LnFt	56.00
	36"	LnFt	70.00
	42"	LnFt	91.00
	48"	LnFt	115.00
	54"	LnFt	160.00
	60"	LnFt	190.00
	Add for Bells	Each	2,000.00
	Add for Wet Ground	LnFt	6.50
	Add for Obstructions - Hand Removal	CuFt	65.00
	Add for Obstructions - Machine Removal	CuFt	12.00
	Add for Reinforced Steel	Ton	1,700.0
	Add for Mobilization		10%
0204.0	**EARTH & WATER RETAINER WORK**		
.1	STEEL SHAFT PILING (Including Bracing) (L&M)		
	Heavy -27# -12"-16' Full Salvage	SqFt	19.50
	40# -15"-16' Full Salvage	SqFt	20.75
	57# -18"-16' Full Salvage	SqFt	22.00
	Add per Foot Longer than 16'	SqFt	2.25
	Add for No Salvage	SqFt	9.00
	Light -12 ga - 18" Black to 8' Full Salvage	SqFt	16.20
	10 ga - 18" Black to 8' Full Salvage	SqFt	17.30
	8 ga - 18" Black to 8' Full Salvage	SqFt	22.00
	Add for No Salvage	SqFt	7.25
	Add for Galvanized	SqFt	30%
.2	WOOD SHEET PILING & BRACING		
	6' Deep	SqFt	6.00
	8' Deep	SqFt	6.25
	10' Deep	SqFt	6.80
	12' Deep	SqFt	7.25
.3	H PILES & WOOD SHEATHING (4" Lbr.)		
	10' Deep	SqFt	22.00
	15' Deep	SqFt	23.50
	20' Deep	SqFt	28.50
	25' Deep	SqFt	32.75
	Deduct for Salvage	SqFt	40%
.4	UNDERPINNING	CuYd	775.00
.5	RIP RAP & LOOSE STONE WALLS		
	Machine Placed	CuYd	25.00
	Hand Placed	CuYd	60.00
	and Grouted	CuYd	150.00
.6	CRIBBING		
	6" Open to 10'	SqFt	24.50
	8" Open	SqFt	32.00
0205.0	**SITE DRAINAGE (Including Building Foundations)**		
.1	DITCHING AND BACKFILL (Open) - See 0202.2		
.2	DEWATERING		
	Pumping -2" Pump	Day	62.00
	3" Pump	Day	95.00
	4" Pump	Day	130.00
	Well Points - Ln.Ft. Header - first month	Month	135.00
	Add for Each Additional Month	Month	75.00
	Add for Attending Time	Hour	50.00

0205.0 SITE DRAINAGE, Cont'd...

.3   CULVERTS & PIPING (No Excavation - See 0202.2)

.31   Corrugated Galvanized Metal

	UNIT	LABOR	MATERIAL
Watertight (Bituminous Coated)			
12" Diameter with Bands & Gaskets	LnFt	4.65	8.90
15"	LnFt	5.00	13.20
18"	LnFt	5.55	16.50
24"	LnFt	6.50	22.00
30"	LnFt	7.80	27.50
36"	LnFt	9.25	34.10
48"	LnFt	18.55	53.35
Non - Watertight			
12" Diameter with Bands & Gaskets	LnFt	3.70	7.50
15"	LnFt	4.50	8.25
18"	LnFt	5.30	11.00
24"	LnFt	7.40	13.90
30"	LnFt	9.40	16.50
36"	LnFt	10.30	27.60
48"	LnFt	10.90	37.95
Add for Oval Arch Shape	LnFt	20%	-
Aprons			
18"	Each	18.40	72.60
24"	Each	21.00	107.80
30"	Each	23.30	176.00
36"	Each	26.50	281.60

.32   Reinforced Concrete Pipe - #3 - Class 5

	UNIT	LABOR	MATERIAL
12" Diameter	LnFt	6.00	12.10
18"	LnFt	8.20	18.70
24"	LnFt	11.70	25.30
27"	LnFt	14.00	38.50
30"	LnFt	17.70	46.20
36"	LnFt	21.70	50.60
48"	LnFt	32.90	79.20
Add for Each Added Foot in Depth	LnFt	5%	-
Add for Apron 36" (Flared End)	Each	180.25	500
Add for Trash Guards (Galvanized)	Each	153.70	850

.33   Vitrified Clay

	UNIT	LABOR	MATERIAL
12"	LnFt	6.40	12.50
24"	LnFt	10.60	46.50

.4   CATCH BASINS & MANHOLES (No Excavation)

	UNIT	LABOR	MATERIAL
48" Cast in Place (Concrete)	LnFt	161.10	162.75
48" Brick or Block	LnFt	156.90	168.00
48" Precast Concrete	LnFt	103.90	97.65
48" Precast Concrete Collar	LnFt	133.60	168.00
Add for Cover - Precast Concrete	Each	73.10	152.25
Add for Base - Precast Concrete	Each	95.40	157.50
Add for Covers & Grates - H.D.C.I.	Each	43.50	178.50
Add for Covers & Grates - L.D.C.I.	Each	43.50	178.50
Add for Covers - H.D. Watertight	Each	43.50	304.50
Add for Steps	Each	-	15.75
Add for Adjusting Rings 2"	Each	15.40	15.75

.5   FOUNDATION DRAINAGE (Subdrainage)

	UNIT	LABOR	MATERIAL
4" Clay Pipe	LnFt	1.30	1.75
4" Corrugated - Perforated	LnFt	1.10	1.10
4" Plastic Pipe - Perforated	LnFt	1.10	1.30
Solid	LnFt	1.10	1.00
6" Clay Pipe	LnFt	1.50	1.95
6" Corrugated - Perforated	LnFt	1.20	1.60
6" Plastic Pipe - Perforated	LnFt	1.20	2.95
Solid	LnFt	1.20	2.45
Add for Porous Surround 2'x 2' ($8.00 CuYd)	LnFt	1.50	2.80

0206.0 PAVEMENT, CURBS AND WALKS

.1   PAVING (PARKING LOTS AND DRIVEWAYS)

		UNIT	COST
.11	Bituminous (Material $185/Ton, 5% Asphalt, 10-Mile Delivery)		
	1 1/2" Wearing 10.5 SqYd/Ton	SqYd	6.00
	2" Wearing 9.0 SqYd/Ton	SqYd	7.50
	2 1/2" Wearing 7.5 SqYd/Ton	SqYd	8.60
	3" Wearing 6.0 SqYd/Ton	SqYd	10.00
	Add for Paths and Small Driveways	SqYd	5.50
	Add for Patching	SqYd	12.40
	Add for Asphalt Content per 1%	SqYd	2.50
	Add for Over 10-mile Delivery - Ton/Mile	SqYd	.40
	Deduct for Base Asphalt	SqYd	10%
	Add for Seal Coat	SqYd	1.90
	Add for Striping - Paint	LnFt	.30
	Add for Striping - Plastic	LnFt	.60
	Base Course for Above - Add:		
	4" Sand & Gravel $8.80 Ton and 1/3 Ton	SqYd	4.70
	4" Crushed Stone $10.90 Ton	SqYd	5.60
	6" Sand & Gravel $7.70 Ton and 1/2 Ton	SqYd	5.70
	6" Crushed Stone $10.00 Ton	SqYd	6.90
	Add per Mile over 10 Miles	SqYd	.40
	Deduct for Belly Dump Delivery - $.75 Ton	SqYd	.40
	4" Sand & Gravel	Ton	10.20
	4" Crushed Stone	Ton	11.40
	6" Sand & Gravel	Ton	9.20
	6" Crushed Stone	Ton	10.70
	Add per Ton Mile Over 10 Miles	Ton	.40
	Deduct for Belly Dump Delivery	Ton	.95
.12	Concrete (4000# Ready Mix, $74 CuYd - Machine Placed)		
	6" Reinforced with 6 x 6, 8-8 Mesh	SqYd	24.15
	7" Reinforced with 6 x 6, 6-6 Mesh	SqYd	25.10
	8" Reinforced with #4 Rods - 12" O.C.	SqYd	28.50
	9" Reinforced with #5 Rods - 12" O.C.	SqYd	31.00
	10" Reinforced with #6 Rods - 8" O.C.	SqYd	32.10
	12" Reinforced with #8 Rods - 8" O.C.	SqYd	40.70
	Add for Base - Same as Base Prices Above		
	Add to Above for Hand Placed	SqYd	25%
	Add for Driveways	SqYd	50%
.13	Stabilized Aggregate (Delivery by Dump Truck)		
	4" Gravel - $8.80 Ton - 3/4" Compacted	SqYd	5.40
	4" Crushed Stone - $10.00 Ton - 3/4" Compacted	SqYd	6.20
	6" Gravel - $8.00 Ton - 3/4" Compacted	SqYd	6.10
	6" Crushed Stone - $9.30 Ton - 3/4" Compacted	SqYd	7.90
	6" Pulverized Concrete	SqYd	8.30
	Add for Soil Cement Treatment	SqYd	1.05
	Add for Oil Penetration Treatment	SqYd	3.30
	Add for Calcium Chloride Treatment	SqYd	.55
	Add per CuYd Mile Over 10 Miles	SqYd	.40
	Deduct for Belly Dump Truck Delivery	SqYd	.45
.14	Brick - 4" x 8" x 2 1/2"	SqYd	84.00
	Add for Sand Bed - Compacted	SqYd	7.10
	Add for Mortar Setting - Bed & Joints	SqYd	7.80

0206.0 PAVEMENT, CURBS AND WALKS, Cont'd...
.2 CURBS & GUTTERS (Cost Incl. Excavation & Equip.)

.21 Concrete - Cast in Place	UNIT	LABOR	MATERIAL
Curb			
6" x 12" Small Job - 200' Day	LnFt	6.20	3.00
6" x 18"	LnFt	7.50	3.55
6" x 24"	LnFt	9.40	4.60
6" x 30"	LnFt	11.00	5.55
Add for Reinforced - Two #5 Rods	LnFt	.75	.70
Add for Gutter - 6" x 12"	LnFt	4.60	2.40
Add for Gutter - 6" x 18"	LnFt	4.90	3.30
Add for Large Job - Over 200' Day	LnFt		10%
Add for Formless 1000' Day Up	LnFt		20%
Add for Curved or Radius Work	LnFt		40%
Curb & Gutter Rolled			
6" x 12" Small Job	LnFt	7.90	4.40
6" x 18"	LnFt	9.20	6.05
6" x 24"	LnFt	11.10	7.80
Deduct for Large Job - Over 200' Day	LnFt	-	10%
Deduct for Formless 1000' Day Up	LnFt	-	25%
Add for Reinforced - Two #5 Rods	LnFt	.70	.80
Add for Curved or Radius Work	LnFt	-	40%
Add for Approaches - 8' x 16' x 6'	SqFt	3.90	2.30
.22 Concrete - Precast & Pinned - 10" x 10'	LnFt	2.10	7.35
9" x 10'	LnFt	2.00	6.80
.23 Asphalt - 6" x 8'	LnFt	1.10	1.50
Add for Curved Work	LnFt	-	30%
.24 Granite - 6" x 16"	LnFt	5.90	25.00
.25 Timbers, Treated - 6" x 6"	LnFt	2.10	3.75
6" x 8"	LnFt	2.80	5.45
.26 Plastic - 6" x 6"	LnFt	1.90	6.20
.3 WALKS AND DRIVEWAYS			
Bituminous - 1 1/2" with 4" base	SqFt	.50	.72
2" with 4" base	SqFt	.55	.75
Concrete - 4" - Broom Finish	SqFt	1.60	1.30
5" - Broom Finish	SqFt	1.70	1.60
6" - Broom Finish	SqFt	1.75	1.65
Add Mesh 6" x 6", 10 - 10	SqFt	.15	.14
Add for Exposed Aggregate - Washed	SqFt	.70	.07
Add for Coloring (Acrylic)	SqFt	.40	.69
Add for Base - 4" Sand or Gravel	SqFt	.30	.10
Precast Block - Colored 1" with 2" Sand Cushion	SqFt	1.50	.89
Colored 2" with 2" Sand Cushion	SqFt	1.70	1.10
Crushed Rock - 4" Compacted	SqFt	.35	.31
Flagstone - 1 1/4" with Sand Bed	SqFt	4.10	8.10
1 1/4" with Mortar Setting Bed	SqFt	5.80	9.10
Brick - 4" x 8" x 2 1/4" with 4" Sand Bed	SqFt	4.10	3.40
4" x 8" x 2 1/4" with Mortar Setting Bed	SqFt	4.80	4.60
Add for Herringbone or Weave Pattern	SqFt	1.40	1.50
Wood - 2" T&G on 6" x 6" Timbers	SqFt	1.60	2.75
2" Boards on 4" x 4" Timbers	SqFt	1.40	2.30
Asphalt Block - 6" x 12" x 3"	SqFt	2.50	5.00
Slate - 1 1/4"	SqFt	6.20	8.50
.4 STAIRS – EXTERIOR	SqFt	7.30	10.00

0207.0 FENCING (L&M) (Ironworkers or Carpenters)

	UNIT	COST
.1 CHAIN LINK (Galvanized 9 ga) Including Intermediate Posts and Concrete Embedded		
5' High - 2" Mesh and 2" O.D. Pipe	LnFt	13.80
6' High	LnFt	14.80
7' High	LnFt	15.40
8' High	LnFt	17.50
10' High	LnFt	32.90
Add for 1 - or 3 - Strand Barbed Wire	LnFt	3.00
Add for Vinyl Coated Wire	LnFt	15%
Add for Aluminum Wire	LnFt	25%
Add for Wood Slats	LnFt	50%
Add for Aluminum Slats	LnFt	75%
Add for Posts - Corner, End and Gate:		
5' High - 3" O.D.	Each	42.40
6' High - 3" O.D.	Each	47.70
7' High - 3" O.D.	Each	60.40
8' High - 3" O.D.	Each	70.00
10' High - 3" O.D.	Each	100.70
Add for Gate Frames:		
5' High	LnFt	42.40
6' High	LnFt	44.50
7' High	LnFt	56.20
8' High	LnFt	70.00
10' High	LnFt	84.80
.2 WELDED WIRE (Galvanized 11 ga) Including Intermediate Posts and Concrete Embedded		
4' High - 2" x 4"	LnFt	11.10
5' High	LnFt	12.50
6' High	LnFt	13.60
Add for Posts:		
4' Long	Each	24.40
5' Long	Each	26.50
6' Long	Each	28.60
Add for Gate Frames:		
4' Long	LnFt	23.30
5' Long	LnFt	28.60
6' Long	LnFt	30.75
Add for Slats	LnFt	5.30
.3 MESH WIRE (Galvanized 14 ga)		
3' High	LnFt	5.60
4' High	LnFt	6.90
5' High	LnFt	7.40
.4 WOOD FENCING		
Rail - 4' Cedar - 3 Boards	LnFt	15.40
Redwood - 3 Boards	LnFt	17.50
Pine, Painted - 3 Boards	LnFt	15.90
Vertical - 6' Pine, Painted	LnFt	19.60
Redwood	LnFt	24.40
Cedar	LnFt	24.40
Weave - 6' Redwood	LnFt	25.40
8' Redwood	LnFt	30.00
Picket - 4' Cedar	LnFt	12.70
6' Cedar	LnFt	15.90
Split Rail - 4' - 3 Rail	LnFt	8.75
Add for Post Set in Concrete	Each	5.30
.5 GUARD RAILS - Corrugated - Galvanized with Wood Posts	LnFt	31.80
Galvanized with Steel Posts	LnFt	33.90
Wire Cable - with Wood Posts, 3 - Strand	LnFt	15.90
with Steel Posts, 3 - Strand	LnFt	16.70
with no posts, 3-strand	LnFt	7.40

0208.0 RECREATIONAL FACILITIES

		UNIT	COST
.1	PLAYING FIELDS AND OUTDOOR COURTS		
.11	Football Fields - Field Size - 300' x 160		
	Minimum Use Size - 306' x 180'		
	Artificial Turf	Each	685,000.00
	Natural Turf	Each	305,000.00
.12	Tennis Courts -Court Size - 78' x 36'		
	Minimum Use Size - 108' x 48'		
	Asphalt with Color Surfacing - 1 Court	Each	25,400.00
	2 Courts	Each	20,700.00
	3 Courts	Each	20,000.00
	Incl. 3" Asphalt, 4" Gravel Base & Striping		
	Synthetic	Each	35,000.00
	Concrete	Each	29,700.00
	Add for Practice Boards	Each	2,700.00
	Add for Net Posts (4" Pipe with Ratchet)	Pair	650.00
	Add for Steel nets	Each	650.00
.13	Track and Field		
	Track - 440 yard - Track Size 21' x 1,320'		
	Synthetic Turf - 1" Polyurethane	Each	148,400.00
	1" Polyurethane	SqYd	42,000.00
	3" Bituminous Base	SqYd	7.95
	6" Gravel Base	SqYd	5.80
	Rubberized Asphalt	Each	69,000.00
	1" with 12% Resiliency	SqYd	11.70
	Add Bituminous Base - same as above	SqYd	7.40
	Cinder	Each	58,000.00
	High Jump - 50' Radius with 12' x 16' P.T.	Each	10,600.00
	Long Jump - 200' x 4' Runway with 20' x 8' P.T.	Each	5,100.00
	Pole Vault - 130' x 4' Runway with 16' x 16' P.T.	Each	3,200.00
.14	Basketball Courts -		
	Concrete - 4" Grade School - Court 42' x 74'	Each	18,000.00
	High School - Court 50' x 84'	Each	21,200.00
	College - Court 50' x 94'	Each	25,000.00
	Asphalt - 1 1/2" - Grade School	Each	8,700.00
	High School	Each	9,800.00
	College	Each	11,800.00
	Add for Back Stops	Each	1,500.00
.15	Volley Ball Courts - Court Size 30' x 60'	Each	10,600.00
	Minimum Use Size - 36' x 66'		
	Concrete - 4"	Each	9,500.00
	Asphalt - 1 1/2"	Each	7,400.00
	Add for Posts and Inserts	Pair	620.00
.16	Shuffleboard Courts - 6' x 52'		
	Concrete - 4"	Each	1,400.00
	Asphalt - 1 1/2"	Each	900.00
.17	Miniature Golf (9 - Hole) Bases - Concrete	Each	10,600.00
	Equipment and Carpet	Each	8,600.00
.18	Horse Shoe Courts - 10' x 40'	Each	600.00
	See Division 11 for Playground Equipment		
.19	Baseball and Softball Fields		
	45' Radius with Backstop	Each	13,800.00
	65' Radius with Backstop	Each	17,500.00

0208.0 RECREATIONAL FACILITIES, Cont'd...

	UNIT	COST
.2 RECREATIONAL EQUIPMENT		
Climbers - Arch	Each	535.00
Dome	Each	630.00
Circular	Each	1,120.00
Catwalks	Each	730.00
Slides - Straight - 6'	Each	960.00
8'	Each	1,230.00
10'	Each	1,820.00
Spiral - 7'	Each	4,400.00
10'	Each	6,600.00
Spring Equipment - Single	Each	420.00
2 Unit	Each	630.00
4 Unit	Each	1,160.00
Swings - 2 Leg - 2 Seat - 8' High	Each	710.00
2 Leg - 4 Seat	Each	1,030.00
3 Leg - 3 Seat - 8' High	Each	790.00
6 Seat	Each	1,200.00
Add for 12' High	-	20%
Add for Nursery Seats	Each	290.00
Add for Saddle Seats	Each	660.00
Add for Tire Swings	Each	170.00
Teeter Totters	Each	1,050.00
Whirls - Small - 6'	Each	870.00
Large - 10'	Each	1,580.00
.3 SITE FURNISHINGS		
Benches - Steel Leg		
Stationary - with Black Wood Slats - 8'	Each	430.00
with Black Aluminum Slats - 8'	Each	570.00
Portable - 8'	Each	275.00
Player - No Back, Wood Seat - 8'	Each	225.00
No Back, Wood Seat - 14'	Each	360.00
Bike Racks		
Permanent - 10'	Each	590.00
20'	Each	910.00
Portable - 10'	Each	330.00
Bleachers, Elevated		
5 - Row x 15' Steel - Seats 50	Each	3,640.00
10 - Row x 15' Steel - Seats 100	Each	6,300.00
15 - Row x 15' Steel - Seats 150	Each	8,880.00
Litter Receptacles		
Pedestal - 24"	Each	330.00
36"	Each	410.00
Stoves	Each	400.00
Tables - Steel Frame - Wood Slat - 8'	Each	430.00
Aluminum Slat - 8'	Each	465.00
All Wood	Each	275.00
All Aluminum	Each	470.00
.4 SHELTERS	SqFt	15.00

0209.0	LANDSCAPING (L&M) (Laborers)	UNIT	COST	UNIT	COST
.1	SEEDING (.80 lb) - Machine	SqYd	.45	Acre	2,000.00
	Hand	SqYd	1.00	Acre	4,900.00
	Add Fine Grading - Hand	SqYd	.90	Acre	4,250.00
	Add for 30 - Day Maintenance	-	-	Acre	1,430.00
	Deduct for Rye and Wild Grasses	-	-	Acre	640.00
.2	SODDING - Flat ($.80 per SqYd)	SqYd	2.60	-	-
	Slope (Pegged)	SqYd	3.20	-	-
	Add Fine Grading - Hand	SqYd	1.00	-	-
	Add for 30 - Day Maintenance	SqYd	.45	-	-
	Remove and Haul Away Old Sod	SqYd	1.20		
.3	DISCING	SqYd	.95	Acre	4,620.00
.4	FERTILIZING	SqYd	.30	Acre	1190.00
				Ton	970.00
.5	TOP SOIL - 4" Hand - Only	SqYd	3.30	Acre	18,200.00
	Machine and Hand	SqYd	2.30	Acre	11,900.00
		CuYd	30.00	-	-
.6	TREES AND SHRUBS				
	Trees - 2"			Each	470.00
	3"			Each	650.00
	4"			Each	880.00
	5"			Each	1110.00
	6"			Each	1,650.00
	Shrubs - Small - 1' to 3'			Each	25.00
	Large - 3' to 5'			Each	46.00
.7	BEDS				
	Wood Chips, Cedar, 2" - $12 CuYd			SqYd	3.10
	Mulch, Redwood Bark - $30 CuYd			SqYd	6.40
	Stone Aggregate - $15 CuYd			SqYd	3.50
	Planting Soil, 24" - $15 CuYd			SqYd	4.00
.8	WOOD CURBS AND WALLS				
	Curbs - 4" x 4" - Fir or Pine Treated			LnFt	2.60
	Cedar			LnFt	3.40
	6" x 6" - Fir or Pine Treated			LnFt	6.80
	Cedar			LnFt	8.50
	Walls - 4" x 4" - Fir or Pine, Incl. Dead Men			SqFt	7.15
	6" x 6" - Fir or Pine, Incl. Dead Men			SqFt	10.60
	Cedar, Including Dead Men			SqFt	16.40
.9	SPRINKLER SYSTEM, WATERED AREA			CSF	31.40
0210.0	SOIL STABILIZATION				
.1	CHEMICAL INTRUSION			CuFt	17.60
	Pressure Grout			Gal	5.50
.2	CONCRETE INTRUSION			CuFt	32.00
.3	VIBRO FLOTATION			-	-
0211.0	SOIL TREATMENT				
.1	PESTICIDE			SqFt	.30
.2	VEGETATION PREVENTATIVE			SqFt	.10
0212.0	RAILROAD WORK				
.1	NEW WORK			LnFt	150.00
.2	REPAIR WORK			LnFt	21.00
0213.0	MARINE WORK				
0214.0	TUNNEL WORK				

0201.0 DEMOLITION

.32   Selective Building Removals -
No Cutting or Disposal Included (see 0201.34 & 0201.35)
Concrete - Hand Work (see 0201.32 for Machine Work)

	UNIT	COST
8" Walls - Reinforced	SqFt	11.20
Non - Reinforced	SqFt	7.90
12" Walls - Reinforced	SqFt	19.90
12" Footings x 24" wide	LnFt	17.50
x 36" wide	LnFt	24.30
16" Footings x 24" wide	LnFt	30.60
6" Structural Slab - Reinforced	SqFt	7.30
8" Structural Slab - Reinforced	SqFt	9.00
4" Slab on Ground - Reinforced	SqFt	3.40
Non - Reinforced	SqFt	2.50
6" Slab on Ground - Reinforced	SqFt	4.00
Non - Reinforced	SqFt	3.00
Stairs - Reinforced	SqFt	12.40
Masonry - Hand Work		
4" Brick or Stone Walls	SqFt	2.30
4" Brick and 8" Backup Block or Tile	SqFt	4.20
4" Block or Tile Partitions	SqFt	2.30
6" Block or Tile Partitions	SqFt	2.40
8" Block or Tile Partitions	SqFt	2.75
12" Block or Tile Partitions	SqFt	3.50
Miscellaneous - Hand Work (Including Loading)		
Acoustical Ceilings - Attached (Including Iron)	SqFt	1.00
Suspended (Including Grid)	SqFt	.70
Asbestos - Pipe	LnFt	77.00
Ceilings and Walls	SqFt	25.50
Columns and Beams	SqFt	64.00
Tile Flooring	SqFt	3.00
Cabinets and Tops	LnFt	17.40
Carpet	SqFt	54.00
Ceramic and Quarry Tile	SqFt	2.00
Doors and Frames - Metal	Each	89.00
Wood	Each	81.20
Drywall Ceilings - Attached	SqFt	1.30
Drywall on Wood or Metal Studs - 2 Sides	SqFt	1.40
Paint Removal - Doors and Windows	SqFt	1.40
Walls	SqFt	1.10
Plaster Ceilings - Attached (Including Iron)	SqFt	1.85
on Wood or Metal Studs	SqFt	1.95
Roofing - Builtup	SqFt	1.75
Shingles - Asphalt and Wood	SqFt	.60
Terrazzo Flooring	SqFt	3.00
Vinyl Flooring	SqFt	.60
Wall Coverings	SqFt	1.10
Windows	Each	58.30
Wood Flooring	SqFt	.60

.33   Site Removals (Including Loading)

4" Concrete Walks - Labor Only (Non-Reinforced)	SqFt	2.50
Machine Only (Non-Reinforced)	SqFt	.80
6" Concrete Drives - Labor Only (Non-Reinforced)	SqFt	3.10
Machine Only (Reinforced)	SqFt	1.10
6" x 18" Concrete Curb - Machine	LnFt	2.10
Curb and Gutter - Machine	LnFt	2.75
2" Asphalt - Machine	SqFt	.50
Fencing - 8' Hand	LnFt	3.00

		UNIT	COST
0202.0	**EARTHWORK**		
.1	GRADING - Hand - 4" - Site	SqFt	.30
	Hand – 4" Building	SqFt	.50
.2	EXCAVATION - Hand - Open - Soft (Sand)	CuYd	38.50
	Medium (Clay)	CuYd	48.40
	Hard (Shale)	CuYd	77.55
	Add for Trench or Pocket		15%
.3	BACK FILL - Hand - Not Compacted (Site Borrow)	CuYd	17.50
.4	BACK FILL - Hand - Compacted (Site Borrow)		
	12" Lifts - Building - No Machine	CuYd	29.00
	With Machine	CuYd	23.00
	18" Lifts - Building - No Machine	CuYd	28.00
	With Machine	CuYd	22.00
0205.0	**DRAINAGE**		
.4	BUILDING FOUNDATION DRAINAGE		
	4" Clay Pipe	LnFt	4.30
	4" Plastic Pipe - Perforated	LnFt	3.40
	6" Clay Pipe	LnFt	4.70
	6" Plastic Pipe - Perforated	LnFt	5.25
	Add for Porous Surround - 2' x 2'	LnFt	6.25
0206.0	**PAVEMENT, CURBS AND WALKS**		
.2	CURBS AND GUTTERS		
.21	Concrete - Cast in Place (Machine Placed)		
	Curb - 6" x 12"	LnFt	13.80
	6" x 18"	LnFt	16.20
	6" x 24"	LnFt	21.50
	6" x 30"	LnFt	24.25
	Curb and Gutter - 6" x 12"	LnFt	17.30
	6" x 18"	LnFt	21.10
	6" x 24"	LnFt	25.00
	Add for Hand Placed	LnFt	8.95
	Add for 2 #5 Reinf. Rods	LnFt	2.05
	Add for Curves and Radius Work	LnFt	40%
.22	Concrete Precast - 6" x 10" x 8"	LnFt	13.00
	6" x 9" x 8"	LnFt	12.00
.23	Bituminous - 6" x 8"	LnFt	4.00
.24	Granite - 6" x 16"	LnFt	40.00
.25	Timbers - Treated - 6" x 6"	LnFt	8.00
	6" x 8"	LnFt	12.00
.26	Plastic - 6" x 6"	LnFt	11.00
.3	WALKS		
	Bituminous - 1 1/2" with 4" Sand Base	SqFt	1.65
	2" with 4" Sand Base	SqFt	1.75
	Concrete - 4" - Broom Finish	SqFt	4.50
	5" - Broom Finish	SqFt	4.85
	6" - Broom Finish	SqFt	5.00
	Add for 6" x 6", 10 - 10 Mesh	SqFt	.42
	Add for 4" Sand Base	SqFt	.60
	Add for Exposed Aggregate	SqFt	1.05
	Crushed Rock - 4"	SqFt	.75
	Brick - 4" - with 2" Sand Cushion	SqFt	10.60
	4" - with 2" Mortar Setting Bed	SqFt	12.80
	Flagstone – 1 1/4" - with 4" Sand Cushion	SqFt	16.80
	1¼" - with 2" Mortar Setting Bed	SqFt	18.70
	Precast Block - 1" Colored with 4" Sand Cushion	SqFt	3.50
	2" Colored with 4" Sand Cushion	SqFt	4.00
	Wood - 2" Boards on 6" x 6" Timbers	SqFt	6.00
	2" Boards on 4" x 4" Timbers	SqFt	5.40
	Slate - 1 1/4" - with 2" Mortar Setting Bed	SqFt	22.00

* See Section 0505 for Permanent Forms of Corrugated Deck and Steel Lath.

0301.0 GENERAL CONCRETE WORK (S) (Cement Finishers & Laborers)

.1   CONCRETE PLACING & VIBRATING

(3500# Concrete with 5.5 sack cement)

		UNIT	LABOR	MATERIAL
.11	Footings - 1 1/2" Aggregate			
	Wall & Pad Types - Truck Chuted	CuYd	14.80	78.00
	Buggies	CuYd	19.00	78.00
	Crane	CuYd	20.10	78.00
.12	Wall & Grade Beams - 3/4" Aggregate			
	At Grade (to 8' deep) -Truck Chuted	CuYd	15.50	80.00
	Buggies	CuYd	20.90	80.00
	Crane	CuYd	20.10	80.00
	At Grade (over 8' deep, poured with trunks) - Truck Chuted	CuYd	16.40	80.00
	Buggies	CuYd	22.30	80.00
	Crane & Hoppers	CuYd	21.20	80.00
	Above Grade (deeper than 8') - Conveyors	CuYd	15.50	80.00
	Ramp & Buggies	CuYd	22.30	80.00
	Crane & Hoppers	CuYd	21.20	80.00
	Climbing Crane	CuYd	20.10	80.00
	Pumping	CuYd	17.00	80.00
.13	Column & Pedestals - 3/4" Aggregate - Buggies	CuYd	24.40	80.00
	Crane	CuYd	20.90	80.00
	Tower & Buggies	CuYd	24.40	80.00
	Fork Lift	CuYd	43.00	80.00
	Climbing Crane	CuYd	22.30	80.00
	Pumping	CuYd	18.00	80.00
.14	Beams - 3/4" Aggregate - Buggies	CuYd	23.30	80.00
	Crane	CuYd	21.20	80.00
	Tower & Buggies	CuYd	23.30	80.00
	Fork Lift	CuYd	42.40	80.00
	Climbing Crane	CuYd	20.80	80.00
	Pumping	CuYd	18.55	80.00
.15	Slabs - 6" Structure - 3/4" Aggregate - Buggies	CuYd	20.10	80.00
	Crane	CuYd	21.20	80.00
	Climbing Crane	CuYd	19.10	80.00
	Pumping	CuYd	18.00	80.00
.16	Stairs & Landings -			
	Structural - 3/4" Aggregate - Buggies	CuYd	23.30	80.00
	Truck Chuted	CuYd	19.10	80.00
	Climbing Crane	CuYd	21.20	80.00
	Pumping	CuYd	19.10	80.00
	Pan Filled - 1/2" Ready Mixed	CuYd	40.50	85.00
	Hand Mixed (Dry)	CuYd	63.10	77.00
.17	Curbs, Platforms & Misc. Small Pours - 1/2" Aggregate Buggies	CuYd	29.13	85.00
	Fork Lift	CuYd	40.40	85.00
.18	Slabs on Ground - 3/4" Aggregate - Truck Chuted	CuYd	15.90	80.00
	Buggies	CuYd	20.80	80.00
	Crane	CuYd	19.70	80.00
	Conveyors	CuYd	16.40	80.00
.19	Slabs-Over Metal Decks or Lath - 1/2" Aggregate Buggies	CuYd	24.40	85.00
	Tower & Buggies	CuYd	26.00	85.00
	Crane	CuYd	22.50	85.00
.20	Toppings - 1/2" Aggregate - Buggies	CuYd	23.50	85.00
	Towers & Buggies	CuYd	26.00	85.00

Conveying Equipment and Operator not included in above costs. See Page 3-4. Also other variations on Page 3-4

0301.0	GENERAL CONCRETE WORK, Cont'd...	UNIT	LABOR	MATERIAL
	Additions to Concrete Placing Work:			
	Add for Ea 500# Concrete above 3500# Concrete	CuYd	-	2.10
	Add or deduct for Sack Cement	CuYd	-	5.15
	Add for Non Corrosive	CuYd	-	8.20
	Add for High Early Cement Concrete	CuYd	-	8.20
	Deduct for 1 1/2" Aggregate	CuYd	-	1.85
	Add for 3/8" or 1/2" Aggregate	CuYd	-	3.35
	Add for Lightweight Aggregate (4000)	CuYd	3.10	16.50
	Add for Heavyweight Aggregate (Granite) (4000#)	CuYd	6.15	11.30
	Deduct for Fly Ash Concrete (3000#)	CuYd	-	2.70
	Add for Exposed Aggregate Mix	CuYd	-	9.90
	Add for Fibre Reinforcement Mix	CuYd	3.60	8.75
	Add for Less than 6 CuYd Delivery	Load	-	25.75
	Add for Hauls beyond 15 Miles	CuYd Mi	-	1.05
	Add for Time after 7 Minutes per Yard	Minute	-	92.70
	Add for Conveying Equipment and Operators:			
	Towers (Approx. 30 CuYd/ Hour)	CuYd	-	6.20
	Mobile Cranes (Approx. 25 CuYd/ Hour)	CuYd	-	7.20
	Tower Cranes (Approx. 20 CuYd/ Hour)	CuYd	-	7.70
	Fork Lifts (Approx. 5 CuYd/ Hour)	CuYd	-	7.20
	Conveyors (Approx. 50 CuYd/ Hour	CuYd	-	7.70
	Pumping (Approx. 35 CuYd/ Hour)	CuYd	-	4.65
	Add for Floors Above Grade - per Floor Placing	CuYd	1.20	-
	Add for Winter Work:			
	Productivity Loss	CuYd	5.00	-
	Heating Water and Aggregate	CuYd	-	6.20
	Calcium Chloride 1%	CuYd	-	2.30
	Insulation Blanket - Slabs (5 uses)	SqFt	.09	.15
	Walls & Beams (3 uses)	SqFt	.12	.10
	Heaters: Without Operators (Including Fuel - Floor Area)	CuYd	.44	1.65
		or SqFt	.04	.15
	Enclosures (3 uses)- Wall Area	CuYd	.95	.40
		or SqFt	.10	.20
.2	FINISHING (Including Screeds and Bulkheads to 6")			
.21	Rough Screeding - Slabs	SqFt	.35	.07
	Stairs	SqFt	.60	.07
.22	Trowel Finishing - Slabs on Ground	SqFt	.50	.07
	Solid and Pan Slabs	SqFt	.50	.07
	Slab over Corrugated	SqFt	.60	.07
	Topping Slabs	SqFt	.50	.07
	Stairs	SqFt	.90	.07
	Curbs and Bases	SqFt	.90	.07
.23	Broom Finishing - Slabs on Ground	SqFt	.45	.07
	Solid and Pan Slabs	SqFt	.50	.07
	Slab over Corrugated	SqFt	.60	.07
	Topping Slabs	SqFt	.50	.07
	Stairs	SqFt	.90	.07
.24	Float Finish - Slabs on Ground	SqFt	.40	.07
	Solid and Pan Slabs	SqFt	.50	.07
	Slab over Corrugated	SqFt	.50	.07
	Stairs	SqFt	.80	.07
.25	Special Finishing			
.251	Patch Walls, 2 Sides Tie Holes & Honeycomb	SqFt	.20	.08
.252	Rub Walls, 1 Side Carborundum for Fins	SqFt	.10	.04
	With Burlap and Grout	SqFt	.50	.08
.253	Level & Top Floors - Trowel Finish 1"	SqFt	.90	.40
	With Epoxy 1/4"	SqFt	1.75	6.20
.254	Exposed Aggregate Slabs - Washed	SqFt	.70	.06
	Retardant	SqFt	.25	.10
	Add for - Seeding	SqFt	.35	.10
	Add to All Finishing Items Above			
	Winter Production Loss and Cost	SqFt	.20	.06
	Sloped Work	SqFt	.15	05
	Heavyweight Aggregates	SqFt	.20	.06

		UNIT	LABOR	MATERIAL
0301.0	**GENERAL CONCRETE WORK, Cont'd...**			
.255	Bushhammer - Green Concrete	SqFt	1.20	.10
	Cured Concrete	SqFt	1.75	.10
	Sand Blast - Light Penetration	SqFt	.85	.10
	Heavy Penetration	SqFt	1.50	.21
.3	SPECIALTIES			
.31	Abrasives - Non-Slip			
	Alo-Grit (.86 lb & 1/4 lb/SqFt)	SqFt	.20	.25
	Carborundum Grits (1.30 lb & 1/4 lb/SqFt)	SqFt	.20	.41
	Strips 3/8" x 1/4"	LnFt	.75	1.40
	Epoxy Coating (50.00 gal) & 50 SqFt/Gal	SqFt	.30	1.55
.32	Admixtures (per CuYd Concrete)			
	Accelerator (.06 oz)	CuYd	-	1.50
	Air Entraining (.04 oz)	CuYd	-	.90
	Densifiers (.07 oz)	CuYd	-	1.85
	Retarders (.05 oz)	CuYd	-	.90
	Water Reducing (.06 oz)	CuYd	-	2.20
.33	Colors			
	Dust on Type:			
	Black, Brown & Red (.50 lb) 50#/100 SqFt	SqFt	.20	.40
	Green and Blue (.60 lb) 50#/100 SqFt	SqFt	.20	.50
	Integral (Top 1"):			
	Black, Brown & Red (1.50 lb) 27#/100 SqFt	SqFt	.40	.70
	Green and Blue (4.25 lb) 27#/100 SqFt	SqFt	.40	1.50
	Full Thickness			
	Black, Brown & Red	CuYd	24.00	124.00
	Green and Blue	CuYd	24.00	165.00
.34	Curing			
	Curing Compounds			
	Resin (10.00 gal) 200 SqFt/gal	SqFt	.08	.06
	Hydrocide Res.Base (15.00 gal) 200 SqFt/gal	SqFt	.07	.08
	Rubber Base (8.00 gal) 200 SqFt/gal	SqFt	.07	.05
	Wax Base (3.00 gal) 200 SqFt/gal	SqFt	.07	.03
	Asphalt Base (4.00 gal) 200 SqFt/gal	SqFt	.07	.03
	Paper	SqFt	.08	.07
	Polyethylene - 4 mil	SqFt	.08	.05
	Water	SqFt	.08	.01
	Burlap	SqFt	1.00	.07
	Curing and Sealing	SqFt	.12	.07
.35	Expansion and Control Joints			
	Asphalt - Fibre - 1/2" x 4"	LnFt	.40	.26
	1/2" x 6"	LnFt	.40	.36
	1/2" x 8"	LnFt	.50	.50
	Polyethylene Foam - 1/2" x 4"	LnFt	.40	.43
	1/2" x 6"	LnFt	.40	.55
	1/2" x 8"	LnFt	.50	.66
	Sponge Rubber - 1/2" x 4"	LnFt	.40	2.30
	1/2" x 6"	LnFt	.40	3.40
	1/2" x 8"	LnFt	.50	4.60
	Paper - Fibre (No Oil) - 1/2" x 4"	LnFt	.40	.30
	1/2" x 6"	LnFt	.40	.40
	1/2" x 8"	LnFt	.50	.52
	Add for Cap	LnFt	.20	.31
	Add for 3/4" Thickness	LnFt	-	30%
	Add for 1" Thickness	LnFt	-	80%

See 0413.5 for Other Expansion Joint Costs.
See 0410.0 Mortars for Cement Prices.

0301.0 GENERAL CONCRETE WORK, Cont'd...

		UNIT	LABOR	MATERIAL
.36	Grouts - Non-Shrink			
	Iron Oxide (.45/lb)			
	Hand Mixed (1 Cem: 1 Sand: 1 Iron Oxide)	CuFt	6.80	28.00
	Per 1"	SqFt	.85	2.30
	Premixed (.45/lb)	CuFt	2.40	40.00
	Aluminum Oxide (.45/lb)	CuFt	4.90	12.00
	Non-Metallic, Premixed (.35/lb)	CuFt	3.60	30.00
.37	Hardeners and Sealers			
	Acrylic Sealer (7.50 gal) 300 SqFt/gal	SqFt	.13	.05
	Epoxy Sealer (25.00 gal) 300 SqFt/gal	SqFt	.13	.15
	Urethane Sealer (9.50 gal) 400 SqFt/gal	SqFt	.13	.05
	Liquid Hardeners (4.85 gal) 200 SqFt/gal	SqFt	.13	.05
.38	Joint Sealers			
	Rubber Asphalt			
	Hot 1/2" x 1/2" Joint (.70 lb)	LnFt	.60	.50
	Cold 1/2" x 1/2" Joint (.54 lb)	LnFt	.42	.59
	Epoxy (20.00 Qt)	LnFt	.61	1.50
.39	Moisture Proofing (Loose Laid for Slabs)			
	Polyethylene 4 Mil	SqFt	.07	.03
	6 Mil	SqFt	.07	.04
	Asphalted Paper	SqFt	.07	.05
.4	EQUIPMENT (Conv., Finishing, Vibrating, etc.)			
	See Divisions I-6A and I-7A			
.5	TESTS			
	Cylinder - 6" x 12" (7-day and 28-day)	Ea	-	41.00
	Add per Pickup	Ea	-	41.00

0302.0 CONCRETE FORM WORK (S) (Carpenters)
(Lumber @ 500.00 MBF & 3/4" BB Plywood @ 1.05 SqFt)
(Costs Include Erection, Stripping, Cleaning, Oiling)

		UNIT	LABOR	MATERIAL
.1	REMOVABLE FORMS (Expendable and Reusable)			
.11	Footings (2-1/2 BdFt/SqFt and 3 Uses)			
	Wall Type - Constant Elevation	SqFt	2.30	.60
	Pad Type	SqFt	2.40	.60
	Add for Volume Elevations and Changes	SqFt	.17	-
	Add for Deep Foundation (8' or more)	SqFt	.17	-
	Add for Each Foot Form Deeper than 12"	SqFt	.11	.05
	Add for Keyway	LnFt	.48	.12
.12	Walls & Grade Beams (2-1/2 BdFt/SqFt & 3 Uses)			
	Straight Walls - 4' High	SqFt	2.40	.72
	Add for Each Foot Higher than 4'	SqFt	.05	.04
	Add for Walls Above Grade	SqFt	.06	-
	Add for Deep Foundation (8' or more)	SqFt	.13	-
	Add for Pilastered Wall, 24' O.C. Total Area	SqFt	.12	.03
	Add for Pilastered Wall, Pilasters Only	SqFt	4.60	.62
	Add for Openings - Opening Area	SqFt	3.07	.78
	Add for Retaining and Battered Type Wall	SqFt	.90	.12
	Add for Curved Wall	SqFt	1.85	.15
	Add for Brick Ledge	LnFt	1.80	.17
	Add for Parapet Wall - Hung	SqFt	2.80	.28
	Add for Pit or Small Trench Walls	SqFt	1.70	.10
	Add for Lined Forms - Hardwood Panels	SqFt	.40	.50
	Gang Formed Walls - Make Up	SqFt	4.20	6.20
	Move	SqFt	.98	.21
	3/4" B-B Plywood - Avg - 5 Ply - 8 Uses @ 1.05	SqFt	-	-
	7 Ply - 8 Uses @ 1.20	SqFt	-	-
	M.D.O. Plywood - Avg. - 20 Uses @ 1.25	SqFt	-	-

0302.0 CONCRETE FORM WORK, Cont'd...

		UNIT	LABOR	MATERIAL
.13	Columns & Pedestals			
	Sq. & Rectangular (2 1/2 BdFt/ SqFt & 3 Uses)			
	8" x 8"	SqFt	4.70	.87
	12" x 12"	SqFt	4.60	.88
	16" x 16"	SqFt	4.60	.89
	20" x 20"	SqFt	4.40	.90
	24" x 24"	SqFt	4.45	.91
	Add per Foot - Work over 10' Floor Heights	SqFt	.37	-
	Add per Floor above 20' Above Grade	SqFt	.15	-
	Deduct for Ganged Formed (8 Uses)	SqFt	1.30	-
	Round - Steel or Fiberglass (Rent and 3 Uses)			
	12"	LnFt	7.20	4.25
	16"	LnFt	7.00	4.70
	20"	LnFt	8.20	5.60
	24"	LnFt	12.00	6.50
	30"	LnFt	15.60	7.90
	36"	LnFt	19.45	9.80
	Add per Foot - Work over 10" Floor Heights	LnFt	.40	.42
	Round-Fibre (6" to 48" x 18' available)			
	8" Cut Length Prices Used	LnFt	5.00	2.40
	12"	LnFt	5.20	4.50
	16"	LnFt	4.60	7.25
	20"	LnFt	7.10	10.00
	24"	LnFt	9.65	13.10
	30"	LnFt	11.20	16.30
	36"	LnFt	13.90	18.50
	Add for Less than 100 Feet - One Size	LnFt	-	20%
	Add for Seamless Fibre	LnFt	-	15%
	Add for Conical Heads to Steel or Fibre	Each	99.00	60.00
	Add for Beam Fittings or Other Openings	Each	110.00	65.00
.14	Beams (12" Floor Heights and 3 Uses Lumber)			
	Spandrel Beams (Includes Shoring)			
	12" x 48" (4 1/2 BdFt/ SqFt)	SqFt	5.10	1.10
	12" x 42" (4 3/4 BdFt/ SqFt)	SqFt	5.10	1.15
	12" x 36" (5 BdFt/ SqFt)	SqFt	5.25	1.20
	12" x 30" (5 1/4 BdFt/ SqFt)	SqFt	5.35	1.20
	8" x 42" (4 3/4 BdFt/ SqFt)	SqFt	5.35	1.15
	8" x 36" (5 1/4 BdFt/ SqFt)	SqFt	5.60	1.20
	Add for Decks and Safety Rail	SqFt	1.10	.40
	Interior Beams			
	16" x 30" (4 BdFt/ SqFt)	SqFt	5.10	1.10
	16" x 24" (4 1/2 BdFt/ SqFt)	SqFt	5.10	1.15
	12" x 30" (4 3/4 BdFt/ SqFt)	SqFt	5.20	1.20
	12" x 24" (5 1/4 BdFt/ SqFt)	SqFt	5.10	1.25
	12" x 16" (6 1/2 BdFt/ SqFt)	SqFt	5.70	1.40
	8" x 24" (5 1/4 BdFt/ SqFt)	SqFt	5.50	1.30
	8" x 16" (6 1/2 BdFt/ SqFt)	SqFt	5.80	1.40
	Add for Beams Carrying Horizontal Shoring	SqFt	.98	.32
	Add per Foot for Floor Heights over 10'	SqFt	.48	.11
	Add per Floor over 20' Above Grade	SqFt	.15	.10
	Add for Splayed Beams	SqFt	1.70	.20
	Add for Inverted Beams	SqFt	2.20	.20
	Add for Mud Sills (3 Uses)	SqFt	.80	.40

		UNIT	LABOR	MATL
0302.0	**CONCRETE FORM WORK, Cont'd...**			
.15	Slabs (12' Floor Heights and 3 Uses Lumber)			
.151	Solid Slabs (Includes Shoring) (Metal Adj. Beams or Joists)			
	Horizontal Shoring Method			
	To 5" thick (2' OC x 10' Span) 2.5 BdFt/ SqFt	SqFt	2.00	.68
	5" - 8" thick (2' OC x 15' Span) 2.8 BdFt/ SqFt	SqFt	2.10	.76
	8" - 11" thick (2' OC x 20' Span) 3.1 BdFt/ SqFt	SqFt	2.30	.82
	Add for Heights above 10' per foot	SqFt	.30	.08
	Vertical Shoring Method			
	(Wood or Metal Posts, Purlins & Joists)			
	To 5" thick - 2.7 BdFt/SqFt - 4' x 5'6" OC	SqFt	2.10	.74
	5" - 8" thick - 3.0 BdFt/SqFt - 4' x 5'0" OC	SqFt	2.30	.80
	8" - 11" thick - 3.3 BdFt/SqFt - 4' x 4'6" OC	SqFt	3.80	.84
	Add for Heights above 10' per foot	SqFt	.25	.08
	Add for Adjustable Hardware	SqFt	-	.02
	Add for Cantilevered Slabs	SqFt	1.40	.16
	Add for Drop Panels - No Edge included	SqFt	.25	.11
	Flying Form method (8 Uses)			
	To 5" Thick	SqFt	1.70	.63
	5" - 8" Thick	SqFt	1.80	.64
	8" - 11" Thick	SqFt	1.90	.65
	Add for Heights above 8' per foot	SqFt	.23	.07
	Add for 2 Uses	SqFt	.21	.26
	Add for 4 Uses	SqFt	.16	.19
.152	Pan Slabs - Cost same as solid slabs above plus pan forming costs below.			
	Based on 3 Uses and lease of approximately 4,000 Square Feet			
	20" Pan 8" + 2"	SqFt	.95	.76
	10" + 2"	SqFt	.98	.78
	12" + 2"	SqFt	1.00	.80
	14" + 2"	SqFt	1.10	.83
	30" Pan 8" + 2 1/2"	SqFt	.90	.72
	10"+2 1/2"	SqFt	.94	.75
	12"+2 1/2"	SqFt	.98	.78
	14"+2 1/2"	SqFt	1.00	.80
	19" x 19" Dome Pan - 24" x 24" Joist Center			
	6" + 2"	SqFt	.91	.75
	8" + 2"	SqFt	.94	.77
	10" + 2"	SqFt	.96	.77
	12" + 2"	SqFt	1.00	.81
	30" x 30" Dome Pan - 36" x 36" Joist Center			
	10" + 2 1/2"	SqFt	1.00	.82
	12" + 2 1/2"	SqFt	1.10	.84
	14" + 2 1/2"	SqFt	1.20	.88
	Add for 2 Uses	SqFt	.06	.11
	Add for 1 Use	SqFt	.07	.24
	Deduct for 4 Uses	SqFt	.06	.07
	Deduct for 5 Uses	SqFt	.05	.08
	Pan Slabs: Based on 5 Uses & Lease of 4,000 SqFt			
	40" x 40" Dome Pan - 16" + 3"	SqFt	1.30	1.10
	48" x 48" Joist Center - 18" + 3"	SqFt	1.40	1.30
	Deduct for 6 Uses	SqFt	.06	.11
	Deduct for 7 Uses	SqFt	.05	.12
	Add to above units if subcontracted		30%	10%

0302.0   CONCRETE FORM WORK, Cont'd...

		UNIT	LABOR	MATL
.16	Stairs and Landing (3 BdFt and 2 Uses)			
	Structural - Contact Area of			
	Soffits, Stringers and Risers	SqFt	4.30	1.30
	On Ground - Contact Area of			
	Stringers and Risers	SqFt	5.79	1.30
.17	Curbs & Platforms (2 BdFt and 2 Uses)	SqFt	4.00	.76
.18	Edge Forms/Bulkheads for Slabs (2 BdFt, 2 Uses)			
	Floor Supported	SqFt	3.90	.76
	Structural	SqFt	5.40	.84

.2   SHORING, SUPPORTS AND ACCESSORIES
(All items in this section are included in SqFt
costs of Section 0302.1. This section is for
cost-keeping & limited analyzing of variables.)

		UNIT	MATL
.21	Horizontal Shoring		
	10' Span - Monthly Rent Cost	Each	1.90
	SqFt Contact Area	SqFt	.42
	15' Span - Monthly Rent Cost	Each	7.20
	SqFt Contact Area	SqFt	.53
	20' Span - Monthly Rent Cost	Each	8.40
	SqFt Contact Area	SqFt	.72
.22	Vertical Shoring (4' OC - 10' High) - Average	SqFt	.32
	Wood Posts with Ellis Shore Hardware		
	Single - W/T Head - New	Each	13.25
	LnFt Beam - 6 Uses	LnFt	.68
	Double - W/T Head & Bracing - New	Each	27.60
	LnFt Beam - 6 Uses	LnFt	1.20
	Metal Posts - Single - Monthly Rent Cost	Each	3.90
	LnFt Beam - 2 Uses per Month	LnFt	.90
	Double - W/T Head & Bracing	LnFt	7.20
	LnFt Beam - 2 Uses per Month	LnFt	1.60
	Tubular Frames - to 10'		
	Mo. Cost - Incl. Head & Base - 10,000#	Each	15.50
	LnFt - Average - 2 Uses per Month	LnFt	1.30
	Deduct for 6,000#	Each	25%
.23	Column Clamps and Steel Strapping		
	Clamps - Monthly Rent Cost - Steel	Each	4.70
	Lock Fast	Each	8.80
	Gang Form	Each	7.70
	Contact Area - 2 Uses per Month	SqFt.	39
	Strapping - 3/4" x .025 (1.10 lb)	LnFt	.30
	3/4" x .023 (1.00 lb)	LnFt	.30
	Contact Area of Forms	SqFt	.20
.24	Wall and Beam Brackets		
	Brackets (2' OC) - Monthly Rent Cost	Each	.66
	New Cost	Each	6.40
	Contact Area - 4 Uses per Month	SqFt	.16
.25	Accessories - Beam Hangers		
	Tyloops, Lags, Washers & Bolts - Heavy	LnFt	.95
	Light	LnFt	.70

0302.0   CONCRETE FORM WORK, Cont'd...

	UNIT	LABOR	MATERIAL
.3 SPECIALTIES			
.31 Nails, Wire and Ties			
Nails 6 d common (50# box)	CWT	-	55.00
8 d common	CWT	-	56.00
16 d common	CWT	-	60.00
8 d box	CWT	-	86.00
16 d box	CWT	-	90.00
9 ga concrete nails (3/4" to 3")	CWT	-	151.00
Add for Coated Nails	CWT	-	60%
Wire - Black Annealed 9 ga (100# Coil)	CWT	-	86.00
16 ga (100# Coil)	CWT	-	83.00
Average Nail and Wire Cost for Formwork	BdFt	-	.02
	or SqFt	-	.03
Ties - Breakback -8" 3000# 5" ends	Each	-	.92
12" 3000# 5" ends	Each	-	1.10
16" 3000# 5" ends	Each	-	1.10
Add for 5000#	Each	-	.15
Add for 8" ends	Each	-	.09
Average Tie Cost for Formwork (2' OC)			
Wall Area	SqFt	-	.40
Contact Area Form	SqFt	-	.22
Gang Form Contact Area	SqFt	-	.23
Coil Bolts - 8"	Each	-	1.40
12"	Each	-	1.60
.32 Anchors and Inserts (including Layout)			
Anchor Bolts (Concrete) -			
1/2" x 12"	Each	1.10	.80
5/8" x 12"	Each	1.20	1.90
3/4" x 12"	Each	1.40	2.90
Ceiling Type Inserts			
Adjustable - 1/2" Bolt Size x 3'	Each	.90	2.40
5/8" Bolt Size x 3'	Each	1.00	2.40
3/4" Bolt Size x 3'	Each	1.00	2.90
Continuous Slotted	Each	.80	1.80
Threaded - 1/2" Bolt Size	Each	.95	4.05
5/8" Bolt Size	Each	1.00	6.70
3/4" Bolt Size	Each	1.10	7.90
Hanger Wire Type - Drive In	Each	.80	.20
Shell	Each	.80	.24
Ferrule Type - 1/2"	Each	.90	1.10
5/8"	Each	.95	1.35
3/4"	Each	.90	1.90
T-Hanger - 4"	Each	.90	1.00
10"	Each	.90	1.50
14"	Each	1.10	2.05
Wall and Beam Type Insert			
Dovetail Slot - 24 ga (No Anchors)	LnFt	.35	.43
22 ga (No Anchors)	LnFt	.44	.86
Flashing Reglects	LnFt	.60	.70
Shelf Angle Inserts - Add for Bolts,			
Washers, and Nuts -5/8"	Each	.90	3.10
3/4"	Each	1.00	3.90

0302.0 CONCRETE FORM WORK, Cont'd...

		UNIT	LABOR	MATERIAL
	Bolts - 5/8" x 2"	Each	.82	1.20
	3/4" x 2"	Each	.83	1.85
	3/4" x 3"	Each	.85	2.20
	Nuts - 5/8"	Each	.30	.30
	3/4"	Each	.34	.40
	Washers	Each	.07	.30
	Threaded Rod - 3/8" x 36" (Galv)	Each	1.10	1.50
	1/2" x 36" (Galv)	Each	1.20	2.80
	5/8" x 36" (Galv)	Each	1.40	4.10
.33	Chamfers, Drips and Reveals (1 Use)			
	Chamfers - 3/4" Metal	LnFt	.28	.56
	Plastic with Tail	LnFt	.30	.60
	without Tail	LnFt	.32	.50
	Wood	LnFt	.28	.20
	Reveals - 3/4" Plastic	LnFt	.40	1.35
	Drips	LnFt	.36	.45
.34	Stair Nosings and Treads			
	Nosings, Curb Bar - Steel - Galvanized	LnFt	1.20	9.70
	Treads 3" x 1/2" - Cast Iron	LnFt	1.40	11.30
	Extruded Aluminum	LnFt	1.40	7.60
	Cast Aluminum	LnFt	1.40	9.70
.35	Water Stops			
	Polyvinyl Chloride - 6" x 3/16"	LnFt	.85	1.50
	6" Centerbulb	LnFt	.90	2.75
	9" x 3/16"	LnFt	1.00	2.20
	9" Centerbulb	LnFt	1.10	3.50
	Add for 3/8"	LnFt	.18	50%
	Rubber (Neoprene) - 6" Centerbulb	LnFt	.93	8.90
	9" Centerbulb	LnFt	1.10	17.00
	Add per Splice	Each	8.50	-
.36	Divider Strips - White Metal, 16 ga x 1 1/4"	LnFt	.70	.97
	1/2" x 1 1/4"	LnFt	.72	1.00
	3/16" x 1 1/4"	LnFt	.71	1.00
	1/4" x 1 1/4"	LnFt	.71	1.70
	Add for Brass 1/8"	LnFt	.13	.90
.37	Tongue and Groove Joint Forms			
	Asphalt - 3 1/2" x 1/8"	LnFt	.77	.80
	5 1/2" x 1/8"	LnFt	.85	.95
	Metal - 3 1/2"	LnFt	.78	.67
	5 1/2"	LnFt	.85	.86
	Wood - 3 1/2"	LnFt	.78	.26
	5 1/2"	LnFt	.85	.41
	Plastic - 3 1/2"	LnFt	.78	.92
	Add for Stakes	Each	.76	.58
.38	Bearing Pads and Shims			
	Vinyl - 1/8" x 6" x 24"	SqFt	.87	2.20
	1/4" x 6" x 24"	SqFt	1.00	4.40
	Neoprene - 1/8" x 36" (70 Duro)	SqFt	.74	5.50
	1/4" x 36"	SqFt	.80	10.00
.4	EQUIPMENT			

See Pages 1 - 6A and 1 - 7A. Average 5% Labor Cost.

Add to Labor Formwork Costs Above for Winter Construction 10% to 25%

DIVISION #3 - CONCRETE

0303.0 REINFORCING STEEL (S&L) (Ironworkers)

		UNIT	LABOR Subcont	MATERIAL Plain	Epoxy
.1	BARS (Based on Local Trucking)				
	1/4" to 5/8" Large Size Job (100 Tons Up)	Ton	490	1050	1,870
	Medium Job (20 - 99 Tons Up)	Ton	550	1090	1,930
	Small Size Job (5 - 19 Tons)	Ton	570	1144	2,050
	3/4" to 1 1/4" Large Job (100 Tons Up)	Ton	500	1050	1,850
	Medium Job (10 - 99 Tons)	Ton	510	1060	1,870
	Small Size Job (5 - 19 Tons)	Ton	540	1100	1,955
	1 1/2" to 2" Average	Ton	530	1080	1,910
	Add For Galvanizing	Ton	65	480	-
	Add for Bending - Light #2 to #5	Ton	-	260	-
	Heavy #6 and Larger	Ton	-	135	-
	Add for Size Extras - Light #2 to #4	Ton	-	156	-
	Heavy #3 and Larger	Ton	-	125	-
	Add for Spirals - Shop Assembled - Hot Rolled	Ton	-	177	-
	Cold Rolled	Ton	-	250	-
	Unassembled - Hot Rolled	Ton	-	150	-
	Cold Rolled	Ton	-	234	-
	Add for Splicing: #11 Bars - Buttweld	Each	35.00	3.10	-
	#14 Bars - Buttweld	Each	47.00	4.70	-
	#11 Bars - Mech. Buttweld	Each	49.00	12.00	-
	#14 Bars - Mech. Buttweld	Each	55.00	13.50	-

			LABOR	MATERIAL	
	BARS (Based on Local Trucking)				
	Tables of Weights & Prices by LnFt -	Ton	-	1,144.00	
	Small job from warehouse stock for local delivery	Lb	-	.42	
	# 2 Bar 1/4" or .250 .167 lb	LnFt	.15	.15	
	# 3 Bar 3/8" or .375 .376 lb	LnFt	.18	.22	
	# 4 Bar 1/2" or .500 .668 lb	LnFt	.24	.28	
	# 5 Bar 5/8" or .625 1.043 lb	LnFt	.35	.35	
	# 6 Bar 3/4" or .750 1.502 lb	LnFt	.47	.45	
	# 7 Bar 7/8" or .875 2.044 lb				
	# 8 Bar 1" or 1.000 2.670 lb				
	# 9 Bar 1 1/8" or 1.128 3.400 lb				
	#10 Bar 1 1/4" or 1.270 4.303 lb				
	#11 Bar 1 3/8" or 1.410 5.313 lb				
	#14 Bar 1 1/2" or 1.693 7.650 lb				
	#18 Bar 2 1/4" or 2.257 13.600 lb				

.2 ACCESSORIES
Additional sizes, but below are close to standard and are included in tonnage costs above

	UNIT	LABOR	MATERIAL
1/4" to 1 1/2" Slab Bolsters	LnFt	-	.34
1/2" to 2" Beam Bolsters - Upper & Lower	LnFt	-	.63
2" to 3" Beam Bolsters - Upper & Lower	LnFt	-	1.00
3" High Chairs - Continuous	LnFt	-	.52
3/4" x 4" to 6" Joist Chairs	Each	-	.50
1/2" Dowel Bar Tubes - Plastic and Metal	Each	.51	.32
Add for Galvanized Accessories	-	-	.18
Add for Plastic Accessories	-	-	.20

.3 POST TENSION STEEL - PRESTRESSED IN FIELD

	UNIT	LABOR	MATERIAL
Ungrouted - 100 kips	Lb	.84	1.85
200 kips	Lb	.78	1.90
Deduct for Added 100 kips	Lb	.02	-
Add for Chairs	Lb	-	.14

3-12

0303.0 REINFORCING STEEL, Cont'd...

		UNIT	LABOR	MATERIAL
.4	WIRE FABRIC (750 SqFt Roll or 120 SqFt Sheet)			
	6" x 6" x 10/10 (W 1.4)	SqFt	.15	.14
	6" x 6" x 8/8 (W 2.1)	SqFt	.16	.19
	6" x 6" x 6/6 (W 2.9)	SqFt	.20	.24
	6" x 6" x 4/4 (W 4.0)	SqFt	.24	.27
	4" x 4" x 4/4 (W 4.0)	SqFt	.30	.38
	4" x 6" x 8/8 (W 2.0)	SqFt	.20	.21
	Add for Epoxy Coated (120 SqFt Sheet)	SqFt	.02	.23

0304.0 SPECIALTY PLACED CONCRETE (No Fees)

		UNIT	MATERIAL
.1	LIFT SLAB CONSTRUCTION		
	Concrete - 6" Slab	SqFt	4.10
	Formwork	SqFt	1.25
	Reinforcing Steel	SqFt	1.40
	Lifting	SqFt	1.40
	Brace and Assemble	SqFt	.90
		TOTAL	9.05
.2	POST TENSIONED OR STRESSED IN PLACE CONCRETE		
	Concrete - 6" Slab and Columns	SqFt	4.80
	Formwork (Assume Flying Forms)	SqFt	5.10
	Post Tensioning and Reinforcing Steel	SqFt	3.40
		TOTAL	13.30
.3	TILT UP CONSTRUCTION (Incl. Concrete Columns)		
	Concrete - 6" Slab	SqFt	4.50
	Formwork	SqFt	3.00
	Reinforcing Steel	SqFt	2.90
	Tilting Up and Bracing	SqFt	1.35
		TOTAL	11.75

		UNIT	1"	2"	3"
.4	PNEUMATICALLY PLACED CONCRETE (Gunite)				
	Walls - 1" Layers with Mesh	SqFt	3.30	5.15	6.30
	Columns - 1" Layers with Mesh	SqFt	6.30	6.90	8.00
	Roofs	SqFt	4.20	4.85	5.60

0305.0 PRECAST PRESTRESSED CONCRETE (L&M)

		Spans to Approximate	UNIT	COST
.1	BEAMS AND COLUMNS			
	Beams	20'	LnFt	100.00
		30'	LnFt	118.00
	Columns - 12" x 12" x 12' with Plates		LnFt	126.00
	16" x 16" x 12' with Plates		LnFt	155.00
.2	DECKS (Flat) - 24", 40", 48", 72" and 96" Wide - 50 lb. Roof Load			
	4"	12'	SqFt	5.25
	6"	18'	SqFt	5.90
	8"	24'	SqFt	6.30
	10"	30'	SqFt	7.00
	12"	50'	SqFt	8.50
.3	DOUBLE T (Flat)			
	8' x 20" - 30" Deep - 60 lb. Roof Load	40'	SqFt	6.80
		50'	SqFt	7.20
		60'	SqFt	8.40
		70'	SqFt	9.20
		80'	SqFt	10.00
	SINGLE T			
	8' x 36" - 48" Deep	60'	SqFt	11.00
		80'	SqFt	11.90
		100'	SqFt	13.40
		120'	SqFt	15.00
	Add for Floor Decks (100# Load)		SqFt	10%
	Add for Sloped Installation to Decks & T's		SqFt	10%
	Add for 2" Topping for Floors - If Required		SqFt	3.50
	Add per Mobilization		Each	3,200.00

		UNIT	COST
0305.0	**PRECAST PRESTRESSED CONCRETE, Cont'd...**		
.4	WALLS		
	Add to Flat Decks 0305.2 Above	SqFt	1.90
	Add to Double T 0305.3 Above	SqFt	2.05
	Add for Core Wall Type	SqFt	1.85
	Add for Insulated Type	SqFt	1.95
0306.0	**PRECAST CONCRETE (M)**		
.1	BEAMS AND COLUMNS- Same as 0305.1		
.2	DECKS- Same as 0305.2		
.3	PANELS OR FACING UNITS (L&M)		
	Insulated - 8' x 16' x 12"		
	Exposed Aggregate		
	Gravel	SqFt	14.80
	Quartz & Marble	SqFt	17.00
	Trowel or Brush		
	Grey Cement	SqFt	14.40
	White Cement	SqFt	15.65
	Deduct for Non-Insulated 8' x 16' x 8"	SqFt	1.30
	Add for Banded	SqFt	.50
.4	PLANK (S)		
	2" x 24" 6' Span	SqFt	4.90
	3" x 24" 8' Span	SqFt	5.50

			LABOR	MATERIAL
.5	SPECIALTIES (S)			
	Curbs 6" x 10" x 8'	LnFt	2.20	7.00
	8" x 10" x 8'	LnFt	2.70	7.70
	Copings 5" x 13"	LnFt	6.00	10.00
	Sills & Stools 5" x 6"	LnFt	4.80	12.60
	Splash Blocks 3" x 36" x 16"	Each	9.75	25.20
	5" x 36" x 16"	Each	12.30	26.70
	8" x 36" x 16"	Each	16.20	31.20
	Medians	LnFt	9.10	40.80

				COST
0307.0	**PRECAST CEMENTITIOUS PLANK (L&M)**			
.1	EXPANDED MINERALS (Includes Clips, Rods & Grout)			
		SPANS TO APPROX.		
	3" x 24"	6'-0"	SqFt	3.80
	4" x 24"	9'-6"	SqFt	4.10
.2	GYPSUM (Metal Edged)			
	2" x 15"	4'-0"	SqFt	3.90
	2" x 15"	7'-0"	SqFt	4.20
.3	WOOD FIBRE			
	2" x 32"	3'-0"	SqFt	3.60
	2 1/2" x 32"	3'-6"	SqFt	3.80
	3" x 32"	4'-0"	SqFt	4.20

			COST
0308.0	**POURED IN PLACE CEMENTITIOUS CONCRETE (L&M)**		
.1	EXPANDED MINERALS - 2"	SqFt	2.40
.2	EXPANDED MINERALS & ASPHALT - 2"	SqFt	2.60
.3	GYPCRETE - 1/4"	SqFt	1.20
	1"	SqFt	1.50

All examples below are representative samples and have great variables because of design efficiencies, size, spans, weather, etc. All figures include a contractor's fee of 10% for Footings and Slabs on Ground (low risk) and a 15% fee for Walls, Columns, Beams and Slabs (high risk). All totals include material prices as follows: Concrete $78.00 CuYd, Lumber $500.00 M.B.F. (3 uses), Plywood $1.05 SqFt, and Reinforcing Steel $1050.00 Ton. Taxes and insurance on Labor added at 35%. Includes General Conditions at 10%. CuYd Costs are for information only--not quick estimating.

	Unit	Unit Cost With ReinSteel	Unit Cost Without ReinSteel	CuYd Cost With ReinSteel
FOOTINGS (Including Hand Excavation)				
Continuous 24" x 12" (3 # 4 Rods)	LnFt	21.10	19.20	285.00
36" x 12" (4 # 4 Rods)	LnFt	27.10	27.30	244.00
20" x 10" (2 # 5 Rods)	LnFt	18.00	17.30	347.00
16" x 8" (2 # 4 Rods)	LnFt	15.30	17.95	458.00
Pad 24" x 24" x 12" (4 # 5 E.W.)	Each	74.55	55.65	504.00
36" x 36" x 14" (6 # 5 E.W.)	Each	149.00	106.05	447.00
48" x 48" x 16" (8 # 5 E.W.)	Each	247.00	205.80	420.00
WALLS (#5 Rods 12" O.C. - 1 Face)				
8" Wall (# 5 12" O.C. E.W.)	SqFt	15.00	12.20	588.00
12" Wall (# 5 12" O.C. E.W.)	SqFt	16.10	14.70	434.00
16" Wall (# 5 12" O.C. E.W.)	SqFt	18.00	16.60	360.00
Add for Steel - 2 Faces - 12" Wall	SqFt	1.90	-	55.50
Add for Pilastered Wall - 24" O.C.	SqFt	.90	-	24.00
Add for Retaining or Battered Type	SqFt	2.70	-	74.00
Add for Curved Walls	SqFt	4.95	-	144.00
COLUMNS				
Sq Cornered 8" x 8" (4 # 8 Rods)	LnFt	35.90	26.00	2,153.00
12" x 12" (6 # 8 Rods)	LnFt	59.30	48.40	1,601.00
16" x 16" (6 # 10 Rods)	LnFt	80.95	60.90	1,213.00
20" x 20" (8 # 20 Rods)	LnFt	111.80	78.10	1,082.00
24" x 24" (10 # 11 Rods)	LnFt	137.55	93.45	929.00
Round 8" (4 # 8 Rods)	LnFt	25.60	17.85	1,811.00
12" (6 # 8 Rods)	LnFt	39.80	25.00	1,292.00
16" (6 # 10 Rods)	LnFt	54.60	35.20	1,092.00
20" (8 # 20 Rods)	LnFt	76.90	51.80	992.00
24" (10 # 11 Rods)	LnFt	113.40	72.25	924.00
BEAMS				
Spandrel 12" x 48" (33 # Rein. Steel)	LnFt	138.60	109.20	935.00
12" x 42" (26 # Rein. Steel)	LnFt	121.30	95.00	930.00
12" x 36" (21 # Rein. Steel)	LnFt	98.30	82.95	884.00
12" x 30" (15 # Rein. Steel)	LnFt	89.25	82.20	964.00
8" x 48" (26 # Rein. Steel)	LnFt	114.45	104.15	1,124.00
8" x 42" (21 # Rein. Steel)	LnFt	109.70	98.20	1,267.00
8" x 36" (16 # Rein. Steel)	LnFt	100.80	84.75	1,360.00
Interior 16" x 30" (24 # Rein. Steel)	LnFt	80.10	73.60	641.00
16" x 24" (20 # Rein. Steel)	LnFt	79.00	69.10	788.00
12" x 30" (17 # Rein. Steel)	LnFt	86.30	69.50	924.00
12" x 24" (14 # Rein. Steel)	LnFt	66.00	58.10	889.00
12" x 16" (12 # Rein. Steel)	LnFt	55.40	45.40	1,103.00
8" x 24" (13 # Rein. Steel)	LnFt	65.10	55.30	1,302.00
8" x 16" (10 # Rein. Steel)	LnFt	49.90	43.50	1,496.00

SLABS

	Unit	Unit Cost With ReinSteel	Unit Cost Without ReinSteel	CuYd Cost With ReinSteel
Solid -				
4" Thick	SqFt	9.45	7.60	770.00
5" Thick	SqFt	10.90	8.40	710.00
6" Thick	SqFt	12.00	8.00	645.00
7" Thick	SqFt	13.50	11.00	505.00
8" Thick	SqFt	15.20	13.00	610.00
Deduct for Post Tensioned Slabs	SqFt	.85	-	-
Pan -				
Joist -20" Pan - 10" x 2"	SqFt	13.10	11.70	-
12" x 2"	SqFt	14.70	11.90	-
30" Pan - 10" x 2 1/2"	SqFt	13.10	11.20	-
12" x 2 1/2"	SqFt	14.70	12.60	-
Dome -19" x 19" - 10" x 2"	SqFt	14.20	11.30	-
12" x 2"	SqFt	14.90	11.40	-
30" x 30" - 10" x 2 1/2"	SqFt	12.90	11.10	-
12" x 2 1/2"	SqFt	14.00	11.20	-

COMBINED COLUMNS, BEAMS AND SLABS

	Unit	Unit Cost With ReinSteel	Unit Cost Without ReinSteel	CuYd Cost With ReinSteel
20' Span	SqFt	19.30	-	-
30' Span	SqFt	20.40	-	-
40' Span	SqFt	21.75	-	-
50' Span	SqFt	23.40	-	-
60' Span	SqFt	27.90	-	-

STAIRS (Including Landing)

	Unit	Unit Cost With ReinSteel	Unit Cost Without ReinSteel	CuYd Cost With ReinSteel
4' Wide 10' Floor Heights 16 Risers	Riser	169.00	-	1,400.00
5' Wide 10' Floor Heights 16 Risers	Riser	205.00	-	1,480.00
6' Wide 10' Floor Heights 16 Risers	Riser	231.00	-	1,430.00

SLABS ON GROUND

	Unit	Unit Cost With Mesh	CuYd Cost With Mesh
4" Concrete Slab (6 6/10 - 10 Mesh - 5 1/2 Sack Concrete, Trowel Finished, Cured & Truck Chuted)	SqFt	3.57	289.00
Add per Inch of Concrete	SqFt	.58	
Add per Sack of Cement	SqFt	.15	
Deduct for Float Finish	SqFt	.08	
Deduct for Brush or Broom Finish	SqFt	.05	
Add for Runway and Buggied Concrete	SqFt	.18	
Add for Vapor Barrier (4 mil)	SqFt	.15	
Add for Sub - Floor Fill (4" sand/gravel)	SqFt	.42	
Add for Change to 6 6/8 - 8 Mesh	SqFt	.08	
Add for Change to 6 6/6 - 6 Mesh	SqFt	.15	
Add for Sloped Slab	SqFt	.21	
Add for Edge Strip (sidewalk area)	SqFt	.23	
Add for ½" Expansion Joint (20' O.C.)	SqFt	.07	
Add for Control Joints (keyed/ dep.)	SqFt	.17	
Add for Control Joints (joint filled)	SqFt	.18	
Add for Control Joints - Saw Cut (20' O.C.)	SqFt	.32	
Add for Floor Hardener (1 coat)	SqFt	.14	
Add for Exposed Aggregate - Washed Added	SqFt	.42	
Retarding Added	SqFt	.38	
Seeding Added	SqFt	.42	
Add for Light Weight Aggregates	SqFt	.42	
Add for Heavy Weight Aggregates	SqFt	.30	
Add for Winter Production Loss/Cost	SqFt	.36	

See 0206.2 for Curbs, Gutters, Walks and other Exterior Concrete
See 0302.53 for Topping and Leveling of Floors

	UNIT	COST
TOPPING SLABS		
2" Concrete	SqFt	2.85
3" Concrete	SqFt	3.60
No Mesh or Hoisting		
PADS & PLATFORMS (Including Form Work & Reinforcing)		
4"	SqFt	6.80
6"	SqFt	8.50
PRECAST CONCRETE ITEMS		
Curbs 6" x 10" x 8"	LnFt	11.55
Sills & Stools 6"	LnFt	26.25
Splash Blocks 3" x 16"	Each	63.00
MISCELLANEOUS ADDITIONS TO ABOVE CONCRETE IF NEEDED		
Abrasives - Carborundum - Grits	SqFt	.86
Strips	LnFt	2.80
Bushhammer - Green Concrete	SqFt	1.80
Cured Concrete	SqFt	2.50
Champers - Plastic 3/4"	LnFt	.95
Wood 3/4"	LnFt	.60
Metal 3/4"	LnFt	.98
Colors Dust On	SqFt	.74
Integral (Top 1")	SqFt	1.34
Control Joints - Asphalt 1/2" x 4"	LnFt	.84
1/2" x 6"	LnFt	.95
PolyFoam 1/2" x 4"	LnFt	1.11
Dovetail Slots 22 Ga	LnFt	1.45
24 Ga	LnFt	.95
Hardeners Acrylic and Urethane	SqFt	.25
Epoxy	SqFt	.34
Joint Sealers Epoxy	LnFt	2.70
Rubber Asphalt	LnFt	1.45
Moisture Proofing - Polyethylene 4 mil	SqFt	.17
6 mil	SqFt	.21
Non-Shrink Grouts - Non - Metallic	CuFt	39.00
Aluminum Oxide	CuFt	28.00
Iron Oxide	CuFt	43.00
Reglets Flashing	LnFt	1.55
Sand Blast - Light	SqFt	1.30
Heavy	SqFt	2.40
Shelf Angle Inserts 5/8"	LnFt	4.70
Stair Nosings Steel - Galvanized	LnFt	6.80
Tongue & Groove Joint Forms - Asphalt 3 1/2"	LnFt	1.90
Wood 3 1/2"	LnFt	1.40
Metal 3 1/2"	LnFt	1.85
Treads - Extruded Aluminum	LnFt	9.35
Cast Iron	LnFt	10.60
Water Stops Center Bulb - Rubber - 6"	LnFt	10.80
9"	LnFt	19.00
Polyethylene - 6"	LnFt	4.20
9"	LnFt	5.10

The unit costs in this Division are priced as being done by masonry or General Contractors. See Section 4A for Total Subcontracted Cost. In general, the pricing is based on Modular Units, and materials F.O.B. job site in truckload lots. The basic labor units do include mortar mixing, low scaffolding, tending, and labor fringe benefits. Cleaning, high scaffold, equipment reinforcements, taxes and insurance, etc. must be added to basic labor and material units. The basis for Labor Unit Costs can be found as follows: as to labor rate mean, Page 4-1, and as to number of units placed per day, Table 4B. Unit Costs are based on an average crew of 4 Bricklayers to 3 Tenders (including Mortar Mixer and Operators) for Brick and Stone Work, and 1 Bricklayer to 1 Tender for Block and Tile Work.

0401.0 BRICK MASONRY (Bricklayers)

			UNIT	LABOR	MATERIAL
.1	FACE BRICK (See Table 4B for Mfg. Size)				(Truckload)
		Nominal Dimensions			
.11	Conventional	8" x 2-2/3" x 4"			
	Running Bond		M pcs	810.00	520.00
	Common Bond (6 Cs. Headers)		M pcs	820.00	520.00
	Stack Bond		M pcs	840.00	520.00
	Dutch & English Bond		M pcs	880.00	520.00
	Flemish		M pcs	880.00	520.00
.12	Economy	8" x 4" x 4"	M pcs	980.00	770.00
.13	Panel (Triple)	8" x 8" x 4"	M pcs	1,590.00	1,720.00
		8" x 16" x 4"	M pcs	2,630.00	3,400.00
.14	Norman	12" x 2-2/3" x 4"	M pcs	950.00	770.00
.15	King	10" x 2-5/8" x 4"	M pcs	880.00	520.00
.16	Norwegian	12" x 3-1/5" x 4"	M pcs	1040.00	870.00
.17	Saxon-Utility 3" wall	12"x 4"x 3" (thin)	M pcs	1,260.00	1,040.00
	4" wall	12"x 4"x 4"	M pcs	1,280.00	1,070.00
	6" wall	12"x 4"x 6"	M pcs	1,530.00	1,520.00
	8" wall	16"x 3-5/8"x 3-5/8"	M pcs	1,730.00	1,950.00
.18	Meridian	12" x 2-1/4 " x 4"	M pcs	800.00	1590.00
.19	Adobe Brick - Mexican	8" x 2-3/8"x 4"	M pcs	870.00	640.00
		12"x 2-3/8"x 4"	M pcs	1000.00	750.00
.2	COATED BRICK (Ceramic Veneer)		M pcs	900.00	1,160.00
.3	COMMON BRICK - Clay		M pcs	650.00	320.00
	Concrete		M pcs	650.00	286.00
.4	FIRE BRICK - Light Duty (domestic)		M pcs	780.00	1,144.00
	Medium Duty		M pcs	790.00	1,820.00
	Heavy Duty		M pcs	820.00	2,392.00
	Add for Fireplaces		M pcs	640.00	-
	Add:Less than Truckload Lot(14M pcs)				15%
	Add:Full Size Brick		M pcs	4%	8%
	Add:Ceramic Coating - to Common Brick		M pcs	15%	80%
	Deduct: 3" thick Brick (if available)		M pcs	4%	7%
	Add:Ea. Addl. 10' in Fl. Hgt. (or floor)		M pcs	3%	-
	Add:Hauling & Unloading if required		M pcs		See Table 4B
	Add:Breakage & Cutting Allowance		-	-	3%
	Add:Piers & Corbels		M pcs	15%	-
	Add:Sills & Soldiers		M pcs	20%	-
	Add:Floor Brick (over 2")		M pcs	10%	5%
	Add:Weave & Herringbone Pattern		M pcs	20%	7%
	Add:Stack bond		M pcs	8%	-
	Add:Circular or Radius Work		M pcs	20%	-
	Add:Rock Faced		M pcs	10%	-
	Add:Two-Face Joint Striking		M pcs	15%	-
	Add:Arches (including Formwork)		M pcs	75%	20%
	Add:Sawing (Special or Excessive)		M pcs	25%	5%
	Add:Winter Work (below 40 degrees)				
	Production Loss		M pcs	10%	-
	Encls - Wall Area - 1 Side & 3 Uses		Sq Ft	.20	.20
	2 Sides & 3 Uses		Sq Ft	.30	.30
	Heaters & Fuel - Wall Area		Sq Ft	.10	.20
	Heat Mortar		Cu Yd	11.00	5.50
	Deduct: Residential Work		-	10%	-

Crew - 4 Bricklayers to 3 Tenders (Incl. Mortar Mixers & Operators).
See Section 0905.7 for Thin Veneer Brick and Stone (under 2").
See Section 0206.2 for Paver Brick for Sitework.

0402.0 CONCRETE BLOCK (Bricklayers)

		UNIT	LABOR	MATERIAL StdWt	LtWt

.1   CONVENTIONAL
.11   Foundation or Not Struck - Deduct .35 from Labor Units below.
.12   Backup or Struck One Face - Deduct .20 from Labor Units below.
.13   <u>Partition or Struck Two Faces Used for Labor Units below:</u>

	UNIT	LABOR	StdWt	LtWt
12" x 8" x 16" Plain	Ea	2.50	1.65	2.20
Double or Single Corner	Ea	2.30	1.75	2.25
Bull Nose (Sgl or Dbl)	Ea	2.70	1.90	2.40
Bond Beam	Ea	2.60	1.80	2.30
Header	Ea	2.60	1.80	2.30
12" x 8" x 8" Half Length Block	Ea	2.45	1.65	2.30
12" x 4" x 16" Half High Block	Ea	2.45	1.70	2.20
12" x 8" x 8" Lintel	Ea	2.70	1.80	2.30
12" x 16" x 8" Lintel	Ea	2.85	1.90	2.40
12" x 8" x 16" Control Joint	Ea	2.70	1.90	2.45
Half Control Joint	Ea	2.45	1.90	2.55
Fire Rated (4 Hr.)	Ea	2.55	1.90	2.55
L Corner	Ea	2.70	1.75	2.25
Solid or Cap	Ea	2.60	2.00	2.50
8" x 8" x 16" Plain	Ea	2.40	1.20	1.60
Double or Single Corner	Ea	2.45	1.30	1.80
Bullnose (Sgl or Dbl)	Ea	2.50	1.50	2.00
8" x 8" x 8" Half Bullnose	Ea	2.40	1.50	2.00
8" x 8" x 16" Bond Beam	Ea	2.50	1.50	1.90
Header	Ea	2.50	1.20	1.90
8" x 8" x 8" Half Length Block	Ea	2.35	1.20	1.65
8" x 8" x 16" Half High Block	Ea	2.35	1.20	1.75
8" x 8" x 8" Lintel	Ea	2.45	1.40	1.85
8" x 16" x 8" Lintel	Ea	2.70	1.40	1.85
8" x 8" x 16" Control Joint	Ea	2.45	1.55	1.95
8" x 8" x 8" Half Control Joint	Ea	2.30	1.55	1.95
8" x 8" x 16" Fire Rated (2 Hr.)	Ea	2.40	1.40	1.80
(4 Hr.)	Ea	2.40	1.55	1.95
Solid or Cap	Ea	2.40	1.40	1.85
6" x8" x 16" Plain	Ea	2.30	1.00	1.40
Bull Nose (Sgl or Dbl)	Ea	2.30	1.35	1.80
6" x8" x 8" Half Bullnose	Ea	2.30	1.35	1.80
6" x8" x 16" Bond Beam or Lintel	Ea	2.30	1.20	1.65
6" x8" x 8" Half Length Block	Ea	2.30	1.05	1.50
6" x4" x 16" Half High Block	Ea	2.30	1.10	1.50
6" x8" x 16" Fire Rated (2 Hr.)	Ea	2.30	1.20	1.60
L Corner	Ea	2.50	1.30	1.70
Solid or Cap	Ea	2.30	1.40	1.80
4" x 8" x 16" Plain	Ea	2.20	.80	1.25
Bull Nose	Ea	2.25	1.10	1.40
Bond Beam or Lintel	Ea	2.25	1.00	1.30
4" x 8" x 8" Half Length Block	Ea	2.15	.85	1.25
4" x 8" x 16" Half High Block	Ea	2.15	.90	1.20
L Corner	Ea	2.45	1.05	1.35
Solid or Cap	Ea	2.30	1.15	1.45

0402.0   CONCRETE BLOCK (Bricklayers)

		UNIT	LABOR	MATERIAL StdWt	LtWt
.13	Partition or Struck Two Faces , Cont'd...				
	16" x 8" x 16" Plain	Ea	3.00	2.00	2.70
	16" x 8" x 8" Half Length	Ea	2.80	1.95	2.68
	14" x 8" x 16" Plain	Ea	2.65	1.95	2.65
	14" x 8" x 8" Half Length	Ea	2.50	1.90	2.63
	14" x 8" x 16" Bond Beam	Ea	2.50	2.20	2.83
	10" x 8" x 16" Plain	Ea	2.45	1.50	2.06
	10" x 8" x 8" Half Length	Ea	2.30	1.48	2.05
	10" x 8" x 16" Bond Beam	Ea	2.50	1.70	2.25
	3" x 8" x 16" Plain	Ea	2.20	.81	1.10
	Solid	Ea	2.30	.85	1.40
	Add or Deduct to all Blockwork Above for:				
	Add: Full Size Block	Ea	.10	.08	-
	Add: Breakage and Cutting	Ea	.10	3%	3%
	Add: Jamb and Sash Block	Ea	.20	.12	.12
	Add: Stack Bond Work	Ea	.20	-	-
	Add: Radius or Circular Work	Ea	1.10	-	-
	Add: Pilaster, Pier or Pedestal Work	Ea	.50	-	-
	Add: Knock-out Block	Ea	.50	.12	.12
	Deduct: Light Weight Block-Installation	Ea	.10	-	-
	Add: Winter Production Loss (below 40°)	Ea	10%	-	-
	Add: Enclosures - Unit	Ea	.20	.09	.09
	Add: Enclosures - Wall Area 1 Side	SqFt	.23	.18	.18
	Add: Enclosures - Wall Area 2 Sides	SqFt	.34	.28	.28
	Add: Heating & Fuel - Wall Area	SqFt	.10	.20	.22
	Deduct: Residential Work	Ea	10%	-	-
	See 0410 for Core Filling for Reinforced Masonry				
	See 0411 for Filling for Insulated Masonry				
.2	ORNAMENTED OR SPECIAL EFFECT				
.21	Shadow wall and Hi-Lite Block				
	Add to Block Prices 0402.1	Ea	.20	.15	.16
.22	Scored Block				
	Add to Block Prices 0402.1 - 1 Score	Ea	.25	.11	.11
	- 2 Score	Ea	.32	.19	.19
.23	Break Off Block				
	Add to Block Prices 0402.1 - Broken	Ea	.21	.68	.68
	- Unbroken	Ea	.30	.62	.62
.24	Split Face Block				
	Add to Block Prices 0402.1	Ea	.21	.29	.31
.25	Rock Faced Block				
	Add to Block Prices 0402.1	Ea	.21	.19	.19
.26	Adobe (or Slump) Block				
	Add to Block Prices 0402.1				
	8" x 8" x 16"	Ea	.24	.33	-
	4" x 8" x 16"	Ea	.23	.23	-
.3	SCREEN WALL - 4" x 12" x 12"	Ea	2.30	1.90	-
.4	INTERLOCKING				
.41	Epoxy Laid				
	Add to Block Prices 0402.1	Ea	-	.16	.16
	Deduct from Labor Costs 0402.1	Ea	.45	-	-
.42	Panelized (No Head Joint Mortar)				
	Add to Block Prices 0402.1	Ea	-	.16	.16
	Deduct from Labor Costs 0402.1	Ea	.53	-	-

0402.0   CONCRETE BLOCK, Cont'd...

		UNIT	LABOR	MATERIAL
.5	SOUND BLOCK			
	Add to Block Prices 0402.1	Ea	.32	.90
	Add for Fillers	Ea	.19	.30
.6	BURNISHED BLOCK - One Face Finish (Truckload Price)			
	Plain 12" x 8" x 16"	Ea	2.86	3.60
	8" x 8" x 16"	Ea	2.70	3.30
	6" x 8" x 16"	Ea	2.60	2.70
	4" x 8" x 16"	Ea	2.50	2.40
	2" x 8" x 16"	Ea	2.40	2.30
	Shapes 12"	Ea	3.35	4.70
	8"	Ea	3.20	3.60
	6"	Ea	3.10	3.00
	4"	Ea	3.00	2.70
	2"	Ea	2.90	2.60
	Add for Bond Beam	Ea	.42	.95
	Add to above for Two Face Finish	Ea	.53	1.80
	Add to above for Scored Face	Ea	.37	.20
	Add for Color	Ea	-	.40
	Add for Bullnose	Ea	.42	.85
.7	PREFACED UNITS - Ceramic Glazed (Truckload Price)			
	12" x 8" x 16" Stretcher	Ea	3.60	6.45
	Glazed 2 Face	Ea	4.20	9.80
	8" x 8" x 16" Stretcher	Ea	3.45	5.95
	Glazed 2 Face	Ea	3.80	9.50
	6" x 8" x 16" Stretcher	Ea	3.50	5.60
	Glazed 2 Face	Ea	3.80	9.40
	4" x 8" x 16" Stretcher	Ea	3.60	5.30
	Glazed 2 Face	Ea	3.90	9.50
	2" x 8" x 16" Stretcher	Ea	3.30	5.30
	4" x 16" x 16" Stretcher	Ea	6.05	18.20
	Add for Scored Block	Ea	.40	.40
	Add for Base Caps, Jambs, Headers	Ea	.67	1.90
	Add for Raked Joints and White Cement	Ea	.80	.20
	Add for Less than Truckload Lot	Ea	-	10%

Crew Ratio - 1 Block Layer to 1 Tender
(Tender includes Mixers & Operators)

Deduct 10% from Block Labor for Residential Construction

0403.0   CLAY BACKING AND PARTITION TILE
(Carload Price)

	UNIT	LABOR	MATERIAL
3" x 12" x 12"	Sq Ft	1.90	1.90
4" x 12" x 12"	Sq Ft	2.00	2.00
6" x 12" x 12"	Sq Ft	2.20	2.40
8" x 12" x 12"	Sq Ft	2.30	2.90

		UNIT	LABOR	MATERIAL
0404.0	**CLAY FACING TILE**			
.1	6T SERIES (SELECT QUALITY - CLEAR & FUNCTIONAL)			
	2" x 5-1/3" x 12" Soap Stretcher (Open Back)	Ea	2.60	3.35
	Soap Stretcher (Solid Back)	Ea	2.65	3.45
	4" x 5-1/3" x 12" 1 Face Stretcher	Ea	2.75	4.65
	Select 2 Face Stretcher	Ea	2.90	5.90
	6" x 5-1/3" x 12" Select 1 Face Bond Stretcher	Ea	2.75	5.10
	8" x 5-1/3" x 12" Select 1 Face Bond Stretcher	Ea	2.90	6.30
	Shapes- Group 1 Square and B/N Corners	Ea	3.00	5.15
	2 Sills and 2" Base	Ea	3.10	5.50
	3 Sills and 4" Base	Ea	3.10	6.75
	4 Starters and Octagons	Ea	3.50	12.50
	5 Radials and End Closures	Ea	3.40	21.00
	Add for Less than Carload Lots	Ea	-	10%
	Add for Designer Colors	Ea	-	20%
	Add for Base Only	Ea	-	40%
.2	8W SERIES (SELECT QUALITY - CLEAR & FUNCTIONAL)			
	2" x 8" x 16" Soap Stretcher (Open Back)	Ea	3.50	5.60
	Soap Stretcher (Solid Back)	Ea	3.60	6.00
	4" x 8" x 16" 1 Face Stretcher	Ea	3.55	6.20
	Select 2 Face Stretcher	Ea	3.80	9.50
	6" x 8" x 16" Select 2 Face 6" Bond Stretcher	Ea	3.70	8.20
	8" x 8" x 16" Select 1 Face 8" Bond Stretcher	Ea	3.90	9.95
	Shapes- Group 1 Square and B/N Corners	Ea	3.80	8.65
	2 Sills and 2" Base	Ea	4.35	11.15
	3 Sills and 4" Base	Ea	4.45	13.85
	4 Starters and Octagons	Ea	4.80	17.00
	5 Radials and End Closures	Ea	4.95	42.00
	Add for Less than Carload Lots	Ea	-	40%
	Add for Designer Colors	Ea	-	20%
	Add for Base Only	Ea	25%	-
0405.0	**GLASS UNITS**			
	4" x 8" x 4"	Ea	3.70	5.50
	6" x 6" x 4"	Ea	4.00	5.60
	8" x 8" x 4"	Ea	4.60	6.50
	12" x 12" x 4"	Ea	5.95	17.85
	Add for Solar Reflective Type	Ea	-	40%
	Accessories: Reinforcing & Expansion Joint	Ea	-	2.20
0406.0	**TERRA-COTTA**			
	Unglazed	SqFt	2.65	6.00
	Glazed	SqFt	3.10	7.20
	Colored Glaze	SqFt	3.20	11.00
0407.0	**MISCELLANEOUS UNITS**			
.1	COPING TILE - 9" Double Slant	LnFt	4.00	9.00
	12" Double Slant	LnFt	4.40	12.00
	Bell	LnFt	3.80	9.90
	Corners & Ends = 4 x Above Material Prices			
.2	FLUE LINING - 8" x 8"	LnFt	3.80	5.20
	8" x 12"	LnFt	4.10	9.40
	12" x 12"	LnFt	4.50	11.00
	16" x 16"	LnFt	7.95	16.70
	18" x 18"	LnFt	9.10	18.70
	20" x 20"	LnFt	15.15	35.40
	24" x 24"	LnFt	20.15	43.00

0408.0 NATURAL STONE (M) (Bricklayers)
(See 0906.0 for Stone under 2")

		SQUARE FOOT		CUBIC FOOT	
.1	CUT STONE	Labor	Matl.	Labor	Matl.
.11	Limestone				
	Indiana (Standard) and Alabama				
	Face Stone 3"	9.00	18.00	36.00	72.00
	4"	10.00	19.55	30.00	58.70
	Sills and Light Trim	-	-	39.00	70.00
	Copings and Heavy Sections	-	-	35.50	68.00
	Add for Select	1.60	-	6.40	
	Minnesota, Texas, Wisconsin, etc.				
	Face Stone 3"	9.00	23.40	36.00	94.00
	4"	10.20	25.50	30.50	76.40
	Sills and Light Trim	-	-	30.60	92.80
	Copings and Heavy Sections	-	-	29.20	90.00
.12	Marble				
	Face Stone 2"	9.20	29.00	55.30	175.00
	3" and 4"	10.00	31.00	38.00	110.00
	Sills and Light Trim	-	-	39.00	114.00
	Copings and Heavy Sections	-	-	35.00	97.00
.13	Granite				
	Face Stone 2"	9.40	26.00	56.60	156.00
	3"	11.40	28.00	45.60	112.00
	4"	13.25	32.00	39.75	97.00
	Sills and Light Trim	-	-	47.20	106.00
	Copings and Heavy Sections	-	-	43.75	102.00
.14	Slate 1 1/2"	10.00	28.00	60.00	168.00
.2	ASHLAR STONE (40 SqFt/Ton - 4" sawed bed)				
.21	Limestone				
	Indiana - Random	9.40	9.70		
	Coursed (2"-5"-8")	9.75	10.90		
	Minn., Ala., Texas, Wisc., etc.)				
	Split Face-1 Size 7 1/2"	10.00	10.70		
	Coursed (2"-5"-8")	11.70	11.75		
	Sawed/Planed Face-Random	9.65	11.90		
	Coursed (2"-5"-8")	9.20	13.10		
.22	Marble- Sawed Random	10.40	19.10		
	Coursed	10.70	19.70		
.23	Granite- Bushhammered-Random	11.00	22.00		
	Coursed	11.40	23.00		
.3	ROUGH STONE (30 s.f. per Ton)				
.31	Flagstone and Rubble 2"	10.40	8.75		
.32	Field Stone or Boulders	9.65	7.60		
.33	Lt.Wt. Boulders (Igneous) (260 s.f. Ton)				
	2" to 5" Veneer (Sawed Back)	8.30	7.60		
	3" to 10" Boulders	9.90	7.50		
.34	Lt.Wt. Limestone (90 s.f. Ton)				
	2" to 4"	9.40	9.40		

0409.0 PRECAST VENEERS AND SIMULATED MASONRY

.1	ARCHITECTURAL PRECAST STONE		
	Limestone and Gravel	8.60	19.00
	Marble	9.40	26.60
	Granite and Quartz	9.50	24.65
.2	PRECAST CONCRETE	8.30	14.70
.3	MOSAIC GRANITE PANELS	11.90	30.60

0410.0 MORTARS

(See Table 4B for Quantities per Unit)

Material Cost used for Units Below (Truckload Prices)

	UNIT	LABOR	MATERIAL
Portland Cement - Gray	Bag- 94#	11.25	9.90
White	Bag- 94#	23.85	19.30
Masonry Cement	Bag- 70#	10.80	7.60
High Early Cement	Bag- 94#	-	11.00
Lime, Hydrated	Bag- 50#	6.75	5.50
Silica Sand	Bag-100#	-	7.30
Fire Clay	Bag-100#	-	21.30
Sand	CuYd	-	21.80
Add for Less than Truckload	-	-	20%
Labor Incl. in Unit Costs: with Mortar Mixer	CuYd	28.00	-
Hand Mixed	CuYd	64.00	-

.1 PORTLAND CEMENT AND LIME MIX
(4.5 Bags Portland and 4.5 Bags Lime)

	UNIT	LABOR	MATERIAL
1:1:6 - Gray - Truckload	CuYd	27.90	94.00
Less than Truckload	CuYd	27.90	112.00
White	CuYd	27.90	133.00
Add for Using Silica Sand	CuYd	23.50	41.00

.2 MASONRY CEMENT MIX (9 Bags Cement)

	UNIT	LABOR	MATERIAL
1:3 - Truckload	CuYd	28.00	90.20
Less than Truckload	CuYd	28.00	100.50

	UNIT	LABOR	MATERIAL
.3 EPOXY CEMENT	CuYd	36.70	185.00
.4 FIRE RESISTANT CEMENT (Clay Mix 300# Mpc)	CuYd	36.70	102.40
.5 ADMIXTURES (Add to Mortar Prices above)			
Waterproofing (1# sack cmt & $ .40/lb)	CuYd	19.00	9.40
Accelerators (1 qt sack cmt & $1.00/lb)	CuYd	19.00	13.10
.6 COLORING (50# CuYd or 90# Mpc & $1.00/lb)	CuYd	21.95	50.30
Standard Brick	M Pcs	26.50	103.20
Add for Heating Mortar Materials	CuYd	13.00	8.00

10% allowed in above Prices for Shrinkage

0411.0 CORE FILLING FOR REINFORCED CONCRETE MASONRY

Based on 3500# Conc.@ $76 CuYd; Mortar @ $90 CuYd; and $26 CuYd to Mix
Deduct 50% from Labor Units if Truck Chuted or Pumped

	Void			Ready Mix		Job Mix	
	CuFt	%	Unit	L	M	L	M
12" x 8" x 16" Plain - 2 Cell	.43	48	Pc	.23	1.28	.40	1.84
	.48	53	SqFt	.27	1.42	.42	2.06
8" x 8" x 16" Plain - 2 Cell	.27	40	Pc	.15	.81	.24	1.14
	.30	44	SqFt	.16	.89	.26	1.30
6" x 8 " x16" Plain - 2 Cell	.19	40	Pc	.11	.57	.36	.81
	.21	44	SqFt	.12	.63	.38	.90
12" x 8" x 16" Bond Beam	.45	50	Pc	.25	1.31	.68	1.90
	.49	55	SqFt	.28	1.42	.80	2.10
8" x 8" x 16" Bond Beam	.21	32	Pc	.12	.61	.19	.90
	.23	35	SqFt	.13	.68	.20	.99
6" x 8" x 16" Bond Beam	.11	25	Pc	.06	.32	.11	.47
	.13	27	SqFt	.07	.38	.12	.55
12" x 16" x 8" Lintel	.52	58	pc	.29	1.52	.47	2.23
	.57	63	SqFt	.32	1.06	.51	2.44
8" x 16" x 8" Lintel	.27	40	Pc	.22	.79	.24	1.15
	.30	44	SqFt	.17	.87	.26	1.28
16" x 8" x 18" Pilaster Block	.96	72	Pc	.53	2.79	.85	4.12
	1.08	81	LnFt	.60	3.12	.95	4.64
Add for Vertical Rein.			Pc	.27	.26	.27	.26
(1 #4 Rods 16" O.C.)			SqFt	.28	.27	.28	.27

0412.0 CORE & CAVITY FILL FOR INSULATED MASONRY

		R Value-No Insulation	R Value-Insulated	UNIT	LABOR	MATERIAL
.1	LOOSE					
.11	Core Fill					
	12" Concrete Block	1.28	3.90	Pc	.36	.44
				SqFt	.39	.47
	Lt. Wt. Concrete Block	2.27	8.30	Pc	.36	.44
				SqFt	.39	.47
	10" Concrete Block	1.20	2.40	Pc	.32	.37
				SqFt	.34	.41
	Lt. Wt. Concrete Block	2.22	5.70	Pc	.31	.38
				SqFt	.34	.43
	8" Concrete Block	1.11	1.93	Pc	.24	.25
				SqFt	.25	.26
	Lt. Wt. Concrete Block	2.18	5.03	Pc	.24	.24
				SqFt	.25	.25
	6" Concrete Block	.90	1.79	Pc	.22	.18
				SqFt	.23	.19
	Lt. Wt. Concrete Block	1.65	2.99	Pc	.22	.18
				SqFt	.23	.19

		R Value	UNIT	LABOR	MATERIAL
	8" Brick	.88	SqFt	.23	.14
	6" Brick	.66	SqFt	.19	.09
	4" Brick	.44	SqFt	.17	.07
	Add to above for Expanded Mica	-	-	10%	200%
	Add to above for Perlite	-	-	-	100%
	10% added for spillage				

			UNIT	LABOR	MATERIAL
.12	Cavity Fill - Per Inch Wall				
	Expanded Styrene	3.8	SqFt	.24	.21
	Fiber Glass	2.2	SqFt	.24	.24
	Rock Wool	2.9	SqFt	.24	.24
	Cellulose	3.7	SqFt	.24	.22
	Add per added inch	-	-	40%	100%
	10% added for spillage				
.2	RIGID				
	Fiber Glass				
	1" - 3# Density	4.35	SqFt	.33	.34
	1.5"	6.52	SqFt	.36	.52
	2"	8.70	SqFt	.40	.68
	Expanded Styrene - Molded (White)				
	1" - 1# Density	4.33	SqFt	.33	.15
	1.5"	6.52	SqFt	.36	.22
	2"	8.70	SqFt	.40	.29
	Add for Embedded Nailer Type	-	-	-	.11
	Expanded Styrene - Extruded (Blue)				
	1" - 2# Density	5.40	SqFt	.33	.39
	1.5"	8.10	SqFt	.36	.18
	2"	10.80	SqFt	.40	.69
	Expanded Urethane				
	1"	7.14	SqFt	.33	.42
	2"	14.28	SqFt	.40	.60
	Perlite				
	1"	2.78	SqFt	.33	.38
	2"	5.56	SqFt	.40	.77
	Add for Glued Applications	-	SqFt	.05	.02

0413.0 ACCESSORIES & SPECIALTIES - MATERIAL

		UNIT	LABOR	MATERIAL
.1	ANCHORS			
	Dovetail-Brick 12 ga Galvanized x 3 - 1/2"	Ea	.20	.46
	Brick 16 ga Galvanized x 3 - 1/2"	Ea	.19	.30
	Brick 16 ga Galvanized x 5"	Ea	.20	.39
	Stone - Stainless	Ea	.20	1.15
	Rigid Wall - 1 - 1/4"x 3/6"x 8" - Galv	Ea	.19	1.02
	Stic - Kips and Clamp	Ea	.24	.21
	Strap Z - 1/8"x 1"x 6" or 8" - Galv	Ea	.40	1.21
	1/8" x 1"x 6" - SS	Ea	.41	2.60
	3/16" x 1 - 1/4" x 8" - Galv	Ea	.42	1.75
	Split Tail - 3/16" x 1 - 1/4" x 6" - Galv	Ea	.40	1.70
	1/8" x 1 - 1/2"x 6" - SS	Ea	.41	3.50
	Wall Clips - 2"x 3"	Ea	.40	.21
	Wall Plugs - Wood Filled	Ea	.40	.23
.2	TIES - WIRE			
	Galv Z - 3/16" Wire - 6" & 8"	Ea	.19	.18
	Zinc Z - 3/16" Wire - 6" & 8"	Ea	.19	.42
	Copperweld Z - 3/16" Wire - 6"	Ea	.19	.47
	3/16" Wire - 8"	Ea	.19	.55
	Adjustable Rect. - 3 /16" Wire 3 - 1/4"	Ea	.19	.32
	3/16" Wire 4 - 3/4"	Ea	.19	.37
	Corrugated-Strap - 26 ga - 7/8" x 7"	Ea	.19	.04
	22 ga - 7/8" x 7"	Ea	.19	.05
	16 ga - 7/8" x 7"	Ea	.21	.08
	Hardware Cloth - 2" x 5" x 1/4" - 23 ga	Ea	.18	.15
	23 ga x 1/4" x 30"	SqFt	.27	.58
.3	REINFORCEMENT			
	#9 Welded Wire - Ladder Type - 4"	LnFt	.18	.16
	6"	LnFt	.18	.17
	8"	LnFt	.19	.19
	10"	LnFt	.19	.20
	12"	LnFt	.21	.21
	16"	LnFt	.23	.29
	Add for 3/16" Wire H.D.	LnFt	-	30%
	Add for Truss Type	LnFt	-	20%
	Add for Rect. Ties - Composite & Cavity	LnFt	.03	.06
	Add for Rect. Adj. Ties - Comp & Cavity	LnFt	.03	.06
	Reinforcement Rods - Horizontal Placement			
	3/8" or #3 .376 lb	LnFt	.19	.22
	1/2" or #4 .668 lb	LnFt	.24	.28
	5/8" or #5 1.043 lb	LnFt	.34	.36
	Add for Vertical Placement	LnFt	25%	-
.4	FIREPLACE AND CHIMNEY ACCESSORIES			
	Ash Dumps and Thimbles - 6" x 8"	Ea	31.00	13.10
	Clean Cut Doors, Cast Iron - 8" x 8"	Ea	29.00	22.00
	12" x 12"	Ea	34.00	46.00
	Dome Dampers, Cast Iron - 24" x 16"	Ea	26.00	53.00
	30" x 16"	Ea	30.00	68.00
	36" x 20"	Ea	45.00	98.00
	42" x 20"	Ea	58.00	103.00
	48" x 24"	Ea	81.50	142.00

0413.0 ACCESSORIES & SPECIALTIES, Cont'd...

		UNIT	LABOR	MATERIAL 1/2"	1"
.5	CONTROL AND EXPANSION JOINT MATERIAL				
	Asphalt - Fibre	SqFt	.59	1.40	2.40
	Polyethylene Foam	SqFt	.60	1.50	2.50
	Sponge Rubber	SqFt	.60	6.80	8.40
	Paper Fibre	SqFt	.58	.95	1.70
	Also in Thk. of 1/8",1/4",3/8" & 3/4"				

		UNIT	LABOR	2 oz	3 oz
.6	FLASHING, THRU WALL				
	Copper Between Polyethylene	SqFt	.85	1.30	1.60
	Rein Paper - 2 oz/1.05	SqFt	.80	1.49	2.23
	Asphalted Fabric - 3 oz	SqFt	.80	2.00	2.73
	Poly Vinyl Chloride - 20 mil	SqFt	.64	-	.25
	30 mil	SqFt	.70	-	.42
	Add for Adhesives			.05	.05

0414.0 CLEANING & POINTING (See 0414 for Restoration)

		UNIT	LABOR		
.1	BRICK & STONE- Acid & Other Chemicals	SqFt	.46	-	.03
	Soap & Water	SqFt	.40	-	-
.2	BLOCK - 1 Face	SqFt	.23	-	-
.3	FACING TILE AND BLOCK				
	Point - White Cement and Sand	SqFt	.80	-	.08
	Silica Sand	SqFt	.98	-	.14
	Add for Scaffold (See 0417)				

0415.0 MASONRY RESTORATION (L&M) (Bricklayers)

		UNIT		Cost
.1	RAKING, FILLING AND TUCKPOINTING			
	Limited - Standard Brick & 3/8" Joint	SqFt		3.20
	Heavy	SqFt		4.00
	All	SqFt		7.80
	Deduct for Stone	SqFt		20%
	Deduct for 4" x 12" Brick	SqFt		30%
	Add for Larger & Smaller Joints	SqFt		25%
.2	SAND BLAST			
.22	Brick -Wet	SqFt		1.80
	Dry	SqFt		1.70
.23	Block -Wet	SqFt		1.70
	Dry	SqFt		1.52
.3	STEAM CLEAN AND CHEMICAL CLEANING	SqFt		1.49
.4	HIGH PRESSURE WATER CLEANING (Hydro)	SqFt		1.05
.5	BRICK REPLACEMENT -Standard	Each		21.00
	4" x 12"	Each		26.25
	Add for Scaffolding	SqFt		.74
	Add for Protection	SqFt		.32
.6	WATER PROOFING - CLEAR	SqFt		.89
	See 0702.3 for Waterproofing Masonry			
	See 0711.0 for Caulking			

0416.0 INSTALLATION OF STEEL & MISC. EMBEDDED ITEMS
See Division 0502

0417.0 EQUIPMENT AND SCAFFOLD

		UNIT	LABOR	MATERIAL
.1	EQUIPMENT AVERAGE OF LABOR			
	See 1-7A for Rental Rates			
.2	SCAFFOLD - INCLUDING PLANK			
	Tubular Frame- to 40' - Exterior	SqFt	.42	.23
	to 16' - Interior	SqFt	.39	.24
	Swing Stage (Incl. Outriggers & Anchorage)	SqFt	.38	.26

The costs below are average but can vary greatly with Quantity, Weather, Area Practices and many other factors. They are priced as total contractor's price with an overhead and fee of 10% included. Units <u>include</u> mortar, ties, cleaning, and reinforcing. <u>Not included</u> are scaffold, hoisting, lintels, accessories, heat, and enclosures. Also added are 27% Insurance on Labor; 10% for General Conditions, Equipment and Tools; and 5% on Material for Sales Taxes. Crew Size is based on 4 Bricklayers to 3 Tenders for Brick and Stone Work, and 4 Bricklayers to 4 Tenders for Block and Tile. See Table 3 for Production Standards. Deduct 5% for Residential Work.

Example:	.1	Face Brick 8" x 2 2/3" x 4"	UNIT	LABOR	MATERIAL	TOTAL
	.11	Running Bond	Each	.76	.50	1.26
		Mortar and Clean Brick	Each	.10	.10	.20
				.86	.60	1.46
		Add: 27% Labor Burden, 5% Sales Tax		.24	.03	.27
				1.10	.63	1.73
		Add: 10% General Conditions & Equipment				.17
						1.90
		Add: 10% Overhead and Fee				.19
		Total Brick Cost			Each Unit	2.09
					or SqFt	14.10

0401.0 BRICK MASONRY

			COST	
.1	FACE BRICK		SqFt	Unit
.11	Conventional (Modular Size)			
	Running Bond - 8" x 2 2/3" x 4"		14.80	2.20
	Common Bond - 6 Course Header		17.33	2.21
	Stack Bond		15.00	2.23
	Dutch & English Bond - Every Other Course Header		23.30	2.36
	Every Course Header		26.50	2.36
	Flemish Bond - Every Other Course Header		17.85	2.36
	Every Course Header		21.26	2.36
	Add: If Scaffold Needed		.80	.13
	Add: For each 10' of Floor Hgt. or Floor (3%)		.42	.05
	Add: Piers and Corbels (15% to Labor)		1.27	.19
	Add: Sills and Soldiers (20% to Labor)		1.50	.23
	Add: Floor Brick (10% to Labor)		.88	.13
	Add: Weave & Herringbone Patterns (20% to Labor)		1.55	.23
	Add: Stack Bond (8% to Labor)		.70	.11
	Add: Circular or Radius Work 20% to Labor)		1.55	.23
	Add: Rock Faced & Slurried Face (10% to Labor)		.88	.13
	Add: Arches (75% to Labor)		4.54	.67
	Add: For Winter Work (below 40°)			
	Production Loss (10% to Labor)		.80	.13
	Enclosures - Wall Area Conventional		.80	.13
	Heat and Fuel - Wall Area Conventional		.32	.04
.12	Econo - 8" x 4" x 3"		12.13	2.70
.13	Panel - 8" x 8" x 4"		10.76	4.80
.14	Norman - 12" x 2 2/3" x 4"		12.02	2.70
.15	King Size - 10" x 2 5/8" x 4"		10.71	2.29
.16	Norwegian - 12" x 3 1/5" x 4"		10.92	2.92
.17	Saxon-Utility - 12" x 4" x 3"		10.46	3.49
	12" x 4" x 4"		10.50	3.52
	12" x 4" x 6"		13.36	4.45
	12" x 4" x 8"		16.17	5.39
.18	Adobe - 12" x 3" x 4"		17.92	2.65
.2	COATED BRICK (Ceramic)		20.97	3.10
.3	COMMON BRICK (Clay, Concrete and Sand Lime)		11.34	1.68
.4	FIRE BRICK -Light Duty		18.70	2.84
	Heavy Duty		22.00	3.30
	Deduct for Residential Work - All Above		-	5%

Example:		UNIT	LABOR	MATERIAL	TOTAL
12" x 8" x 16" Concrete Block - Partition		Each	2.40	1.57	3.97
Mortar (Labor in Block Placing Above)		Each	-	.22	.22
Reinforcing Wire 16" O.C. Horizontal		Each	.10	.12	.22
Clean Block - 2 Faces		Each	.35	-	.35
			2.85	1.91	4.76
Add: 27% Taxes and Insurance on Labor					.77
Add: 5% Sales Tax on Material					.10
					5.63
Add: 10% General Cond., Equip. and Tools					.57
					6.20
Add: 10% Overhead and Profit					.62
Total Block Cost				Each	6.82
				or SqFt	7.67

<u>Not Included</u> - Scaffold, Hoisting Equipment, Lintels, Embedded Accessories, Insulation, Heat and Enclosures. (See "Adds" Below.)

0402.0 CONCRETE BLOCK

		Std Wt Cost		Lt Wt Cost	
.1	CONVENTIONAL (Struck 2 Sides - Partitions)	SqFt	Unit	SqFt	Unit
	12" x 8"x 16" Plain	8.05	7.16	8.82	7.85
	Bond Beam (with Fill-Reinforcing)	10.72	9.53	11.30	10.03
	12" x 8"x 8" Half Block	8.03	7.14	7.61	6.77
	Double End - Header	8.26	7.35	8.86	7.88
	8" x 8" x 16" Plain	7.10	6.30	7.67	6.83
	Bond Beam (with Fill-Reinforcing)	8.58	7.63	9.05	8.05
	8" x 8" x 8" Half Block	6.96	6.20	7.04	6.25
	Double End - Header	6.32	6.51	7.95	7.07
	6 "x 8" x 16" Plain	6.56	5.83	7.15	6.36
	Bond Beam (with Fill-Reinforcing)	7.45	6.64	7.94	7.06
	Half Block	6.52	5.80	7.06	6.28
	4" x 8" x 16" Plain	6.38	5.67	5.88	5.25
	Half Block	5.62	5.00	6.21	5.52
	16" x 8" x 16" Plain	9.14	8.13	10.00	8.87
	Half Block	8.83	7.85	9.66	8.59
	14" x 8" x 16" Plain	8.72	7.77	9.77	8.68
	Half Block	8.26	7.35	8.64	7.69
	10" x 8" x 16" Plain	7.60	6.72	8.34	7.41
	Bond Beam (with Fill-Reinforcing)	8.67	7.72	9.92	8.82
	Half Block	7.27	6.46	7.85	6.98
	Deduct: Block Struck or Cleaned One Side	.26	.24	.26	.24
	Deduct: Block Not Struck or Cleaned Two Sides	.42	.37	.49	.37
	Deduct: Clean One Side Only	.19	.18	.20	.18
	Deduct: Lt. Wt. Block (Labor Only)	.17	.16	.17	.16
	Add: Jamb and Sash Block	.44	.40	.44	.40
	Add: Bullnose Block	.66	.59	.66	.59
	Add: Pilaster, Pier, Pedestal Work	.72	.65	.72	.65
	Add: Stack Bond Work	.28	.25	.28	.25
	Add: Radius or Circular Work	1.30	1.18	1.30	1.18
	Add: If Scaffold Needed	.75	.66	.75	.66
	Add: Winter Production Cost (10% Labor)	.42	.42	.42	.38
	Add: Winter Enclosing - Wall Area	.42	.42	.42	.38
	Add: Winter Heating - Wall Area	.39	.39	.39	.35
	Add: Core & Beam Filling - See 0411 (Page 4A-18)				
	Add: Insulation - See 0412 (Page 4A-18)				

Deduct for Residential Work - 5%

0402.0	**CONCRETE BLOCK, Cont'd...**	Std Wt Cost		Lt Wt Cost	
		SqFt	Unit	SqFt	Unit
.2	ORNAMENTAL OR SPECIAL EFFECT -				
	Add to Units Above in 0402.1				
	Shadow Wall	.47	.42	.49	.44
	Scored Block	.54	.47	.54	.48
	Breakoff	1.32	1.18	1.20	1.07
	Split Face	.77	.68	.82	.72
	Rock Faced	.75	.66	.80	.71
	Adobe (Slump)	.80	.72	-	-
.3	SCREEN WALL - 4" x 12" x 12"	6.10	5.40	-	-
.4	INTERLOCKING - Deduct from Units 0402.1				
	Epoxy Laid	.51	.46	.47	.46
	Panelized	.76	.67	.70	.67
.5	SOUND BLOCK - Add to Block Prices 0402.1			1.79	1.60
.6	BURNISHED - 12" x 8" x 16"			10.98	9.80
	8" x 8" x 16"			10.22	9.10
	6" x 8" x 16"			9.21	8.20
	4" x 8" x 16"			8.55	7.60
	2" x 8" x 16"			8.10	7.20
	Add for Shapes			2.40	2.10
	Add for 2 Faced Finish			3.50	3.10
	Add for Scored Finish - 1 Face			.82	.75
.7	PREFACED UNITS -12" x 8" x 16" Stretcher			16.10	14.30
	(Ceramic Glazed) 12" x 8" x 16" Glazed 2 Face			21.90	19.50
	8" x 8" x 16" Stretcher			15.00	13.35
	Glazed 2 Face			20.70	18.40
	6" x 8" x 16" Stretcher			13.20	11.80
	Glazed 2 Face			20.00	17.80
	4" x 8" x 16" Stretcher			14.20	12.60
	Glazed 2 Face			19.60	17.40
	2" x 8" x 16" Stretcher			13.80	12.30
	4" x 16" x 16" Stretcher			36.10	32.10
	Add for Base, Caps, Jambs, Headers, Lintels			4.01	3.60
	Add for Scored Block			1.20	1.05

0403.0	**CLAY BACKING AND PARTITION TILE**	COST	
		SqFt	Unit
	3" x 12" x 12"	5.32	5.32
	4" x 12" x 12"	5.70	5.70
	6" x 12" x 12"	6.70	6.70
	8" x 12" x 12"	7.45	7.45

0404.0	**CLAY FACING TILE (GLAZED)**		
.1	6T or 5" x 12" - SERIES		
	2" x 5 1/3" x 12" Soap Stretcher (Solid Back)	19.40	8.60
	4" x 5 1/2" x 12" 1 Face Stretcher	20.60	9.15
	2 Face Stretcher	24.40	10.80
	6" x 5 1/3" x 12" 1 Face Stretcher	25.50	11.35
	8" x 5 1/3" x 12" 1 Face Stretcher	28.60	12.70
.2	8W or 8" x 16" - SERIES		
	2" x 8" x 16" Soap Stretcher (Solid Back)	14.80	13.10
	4" x 8" x 16" 1 Face Stretcher	17.30	13.55
	2 Face Stretcher	20.10	17.75
	6" x 8" x 16" 1 Face Stretcher	18.70	16.55
	8" x 8" x 16" 1 Face Stretcher	19.10	16.90
	Add for Shapes (Average)	50%	50%
	Add for Designer Colors	-	20%
	Add for Less than Truckload Lots	-	10%
	Add for Base Only	-	25%

		COST	
		SqFt	Unit
0405.0	**GLASS UNITS**		
	4" x 8" x 4"	49.10	12.80
	6" x 6" x 4"	53.30	13.30
	8" x 8" x 4"	33.10	14.75
	12" x 12" x 4"	30.90	30.90
0406.0	**TERRA-COTTA**		
	Unglazed	11.00	11.00
	Glazed	15.50	15.50
	Colored Glazed	18.20	18.20

			Ln Ft
0407.0	**FLUE LINING**		
	8" x 12"		18.95
	12" x 12"		20.20
	16" x 16"		32.30
	18" x 18"		33.90
	20" x 20"		65.60
	24" x 24"		84.30

			Sq Ft	Cu Ft
0408.0	**NATURAL STONE**			
.1	CUT STONE			
	Limestone- Indiana and Alabama - 3"		37.20	148.70
	4"		40.20	120.65
	Minnesota, Wisc, Texas, etc.- 3"		42.50	170.00
	4"		48.70	146.20
.12	Marble - 2"		52.30	313.75
	3"		54.10	236.30
.13	Granite - 2"		49.10	294.50
	3"		52.90	211.70
.14	Slate - 1 1/2"		49.10	294.80
.2	ASHLAR- 4" Sawed Bed			
.21	Limestone- Indiana - Random		26.80	80.30
	Coursed - 2" - 5" - 8"		28.10	84.40
	Minnesota, Alabama, Wisconsin, etc.			
	Split Face - Random		29.00	86.95
	Coursed		32.00	96.10
	Sawed or Planed Face - Random		30.45	91.35
	Coursed		32.10	96.40
.22	Marble - Sawed - Random		40.22	120.65
	Coursed		41.40	124.10
.23	Granite - Bushhammered - Random		42.50	127.60
	Coursed		45.40	136.10
.24	Quartzite		30.45	91.35
.3	ROUGH STONE			
.31	Rubble and Flagstone		28.10	84.40
.32	Field Stone or Boulders		27.90	83.80
.33	Light Weight Boulders (Igneous)			
	2" to 4" Veneer - Sawed Back		23.40	-
	3" to 10" Boulders		24.60	-
.34	Light Weight Limestone		25.90	-

		Unit	Cost
0409.0	**PRECAST VENEERS AND SIMULATED MASONRY**		
.1	ARCHITECTURAL PRECAST STONE- Limestone	SqFt	36.00
	Marble	SqFt	42.50
.2	PRECAST CONCRETE	SqFt	28.25
.3	MOSAIC GRANITE PANELS	SqFt	40.75
	See 0904 for Veneer Stones (under 2")		
0410.0	**MORTAR**		
	Portland Cement and Lime Mortar - Labor in Unit Costs	CuYd	107.00
	Masonry Cement Mortar - Labor in Unit Costs	CuYd	95.00
	Mortar for Standard Brick - Material Only	SqFt	.42

0411.0 CORE FILLING FOR REINFORCED CONCRETE BLOCK

	COST	
	SqFt	Unit
Add to Block Prices in 0402.0 (Job Mixed)		
12" x 8" x 16" Plain (includes #4 Rod 16" O.C. Vertical)	3.54	2.90
Bond Beam (includes 2 #4 Rods)	3.84	3.42
8" x 8" x 6" Plain (includes #4 Rod 16" O.C. Vertical)	1.98	1.76
Bond Beam (includes 2 #4 Rods)	1.60	1.41
6" x 8" x 16" Plain	1.74	1.55
Bond Beam (includes 1 #4 Rod)	.90	.79
10" x 8" x 16" Bond Beam (includes 2 #4 Rods)	3.44	3.10
14" x 8" x 16" Bond Beam (includes 2 #4 Rods)	4.00	3.57

0412.0 CORE AND CAVITY FILL FOR INSULATED MASONRY

See Page 4-11 for R Values

	SqFt	Unit
.1 LOOSE		
Core Fill - Expanded Styrene		
12" x 8" x 16" Concrete Block	1.26	1.12
10" x 8" x 16" Concrete Block	1.10	.99
8" x 8" x 16" Concrete Block	.80	.71
6" x 8" x 16" Concrete Block	.67	.60
8"Brick - Jumbo - Thru the Wall	.59	.53
6"Brick - Jumbo - Thru the Wall	.47	.42
4"Brick - Jumbo	.40	.36
Cavity Fill - per Inch		
Expanded Styrene	.53	-
Mica	.90	-
Fiber Glass	.55	-
Rock Wool	.55	-
Cellulose	.53	-
.2 RIGID - Fiber Glass - 3# Density, 1"	.95	-
2"	1.47	-
Expanded Styrene - Molded - 1# Density, 1"	.68	-
2"	1.00	-
Extruded - 2# Density, 1"	.97	-
2"	1.47	-
Expanded Urethane - 1"	1.07	-
2"	1.39	-
Perlite - 1"	1.03	-
2"	1.58	-
Add for Embedded Water Type	.19	-
Add for Glued Applications	.06	-

0413.0 ACCESSORIES AND SPECIALTIES

See Page 4-12

0414.0 CLEANING AND POINTING

	SqFt	Unit
Brick and Stone - Acid or Chemicals	.76	-
Soap and Water	.70	-
Block & Facing Tile - 1 Face (Incl. Hollow Metal Frames)	.40	-
Point with White Cement	1.47	-

0415.0 MASONRY RESTORATION - See 0415.0
0416.0 INSTALLATION OF STEEL - See 0502.0
0417.0 SCAFFOLD

	SqFt	Unit
.1 EQUIPMENT, TOOLS AND BLADES		
Percentage of Labor as an average	-	7%
See 1 - 7A for Rental Rates & New Costs		
.2 SCAFFOLD- Tubular Frame - to 40 feet - Exterior	.95	-
Tubular Frame - to 16 feet - Interior	.90	-
Swing Stage - 40 feet and up	1.00	-

TABLE B

	Modular Manufactured Size - Inches			Mortar CuFt	Unit Per	Units Placed	Haul & Unload	Unit
	l	h	d	M	SqFt	ManDay	M PCs	Wt.

0401.0 BRICK MASONRY

.1	Face Brick							
.11	Conventional							
	Running Bond	7-5/8 x 2-1/4 x 3-5/8		13	6.75	600	16.00	4.5
	Common Bond							
	Header Every 6th Course				ƒ1/6	570	-	-
	Header Every 7th Course				ƒ1/7	570	-	-
	Dutch and English Bond							
	Header Every Other Course				ƒ1/2	510	-	-
	Header Every Course				ƒ2/3	510	-	-
	Flemish Bond							
	Header Every Other Course				ƒ1/8	500	-	-
	Header Every Course				ƒ1/3	500	-	-
.12	Econo	7-5/8 x 3-5/8 x 3-5/8		4.50	430	20.00	6.0	
.13	Panel	7-5/8 x 7-5/8 x 3-5/8		2.25	280	40.00	13.5	
		7-5/8 x 15-5/8 x 3-5/8		1.25	180	80.00	27.0	
.14	Norman	11-5/8 x 2-1/4 x 3-5/8		16	4.50	450	20.00	6.0
.15	King	9-5/8 x 3-1/2 x 3-5/8		16	4.70	440	22.00	4.5
.16	Norwegian	11-5/8 x 2-13/16 x 3-5/8		16	3.75	370	22.00	8.0
.17	Saxon 4"	11-5/8 x 3-5/8 x 3-5/8		19	3.00	320	26.00	9.0
	6"	11-5/8 x 3-5/8 x 5-5/8		28	3.00	260	35.00	13.5
	8"	11-5/8 x 3-5/8 x 7-5/8		38	3.00	200	50.00	18.0

Deduct from above dimensions for 1/2" Joint 1/8" x 1/8" x 1/8".
Add to above dimensions for full size 3/8" x 0" x 2/8".

.2	Coated Brick (Glazed)			12	6.75	510	17.00	4.5
.3	Common Brick							
	Clay	7-14 x 2-1/16 x 3-1/2		13	6.75	660	15.00	4.5
	Concrete			13	6.75	660	15.00	4.5
.4	Fire Brick	8-0 x 2-1/4 x 3-1/2		13	6.65	660	15.00	7.0

0402.0 STANDARD AND LIGHT-WEIGHT CONCRETE BLOCK

12" Plain	11-5/8 x 7-5/8 x 15-5/8	65	1.125	190	110.00	5.0	
1/2 Height	11-5/8 x 3-5/8 x 15-5/8		2.25		Lt Wt	38.0	
1/2 Length	11-5/8 x 7-5/8 x 7-5/8		2.25				
8" Plain	7-5/8 x 7-5/8 x 15-5/8	55	1.125	200	80.00	36.0	
1/2 Height	7-5/8 x 3-5/8 x 15-5/8		2.25		Lt Wt	27.0	
1/2 Length	7-5/8 x 7-5/8 x 7-5/8		2.25				
6" Plain	5-5/8 x 7-5/8 x 15-5/8	45	1.125	210	60.00	28.0	
1/2 Height	5-5/8 x 3-5/8 x 15-5/8		2.25		Lt Wt	21.0	
1/2 Length	5-5/8 x 7-5/8 x 7-5/8		2.25				
4" Plain	3-5/8 x 7-5/8 x 15-5/8	35	1.125	210	50.00	24.0	
1/2 Height	3-5/8 x 3-5/8 x 15-5/8		2.25		Lt Wt	18.0	
1/2 Length	3-5/8 x 7-5/8 x 7-5/8		2.25				
16" Plain	15-5/8 x 7-5/8 x 15-5/8	70	2.25	170	150.00	61.0	
14" Plain	13-5/8 x 7-5/8 x 15-5/8	65	2.25	180	90.00	56.0	
10" Plain	9-5/8 x 7-5/8 x 15-5/8	60	2.25	195	85.00	43.0	

Average 2 cubic yards mortar per M Block.

TABLE B

		Modular Manufactured Size - Inches			Mortar CuFt	Unit Per	Units Placed	Haul & Unload	Unit
		l	h	d	M	SqFt	ManDay	M PCs	Wt.
0403.0	**CLAY BACKING & PARTITION TILE**								
	3"	3 x	11-1/2 x	11-1/2	18.7	1.00	260	25.0	13.0
	4"	4 x	11-1/2 x	11-1/2	22.2	1.00	250	29.0	15.0
	6"	6 x	11-1/2 x	11-1/2	16.0	1.00	230	38.0	20.0
0404.0	**CLAY FACING TILE (GLAZED)**								
	2" 6T	1-3/4 x	5-1/16 x	7-3/4	13	2.25	180	20.00	6.5
	4" 6T	3-3/4 x	5-1/16 x	11-3/4		2.25		30.00	10.5
	6" 6T	5-3/4 x	5-1/16 x	11-3/4		2.25		45.00	15.2
	8" 6T	7-3/4 x	5-1/16 x	11-3/4		2.25		55.00	19.0
	2" 8W	1-3/4 x	7-3/4 x	15-3/4	22	1.13	125	40.00	14.0
	4" 8W	3-3/4 x	7-3/4 x	15-3/4		1.13		60.00	21.0
	6" 8W	5-3/4 x	7-3/4 x	15-3/4		1.13		85.00	31.0
0405.0	**GLASS UNITS**								
	6" x 6"	5-3/4 x	5-3/4 x	3-7/8	12.5	4.00	140	18.00	4.0
	6" x 8"	7-3/4 x	7-3/4 x	3-7/8	16.0	2.25	125	26.00	7.0
	12" x 12"	11-3/4 x	11-3/4 x	3-7/8	23.3	1.00	90	33.00	16.0
0406.0	**TERRA-COTTA**								
	3"	3" x	12" x	12"	25	1.00		36.00	10.0
	4"	4" x	12" x	12"	30	1.00		40.00	13.0
	6"	6" x	12" x	12"	40	1.00		47.00	18.0
	8"	8" x	12" x	12"	60				
0407.0	**MISCELLANEOUS UNITS**								
	Coping Tile	9"			50	-	160	35.00	-
		12"			60	-	140	40.00	-
	Flue Lining	8" x	8"		50	-	160	35.00	-
		12" x	12"		55	-	130	45.00	-
		18" x	18"		65	-	60	55.00	-
		21" x	21"		75	-	40	70.00	-
		24" x	24"		90	-	30	90.00	-

		Mortar SqFt	Unit Per	Units Placed ManDay	Haul & Unload M PCs	Unit CuFt
0408.0	**NATURAL STONE**					
	Cut Stone					
	Lime Stone	10.0	-	18	25.00	150.0
	Marble	10.0	-	13	25.00	160.0
	Granite	10.0	-	11	25.00	165.0
	Ashlar Stone	15.0	-	26	25.00	150.0
	Rough Stone - Flagstone	15.0	-	22	25.00	140.0
	Field Stone	15.0	-	22	25.00	160.0
	Light Weight Stone	15.0	-	24	25.00	220.0
0409.0	**PRECAST VENEERS**					
	Architectural Precast Stone	10.0	-	18	25.00	150.0
	Terra-Cotta	10.0	-	17	25.00	130.0
	Precast Concrete	10.0	-	24	25.00	150.0

0501.0 STRUCTURAL METAL (L+M) (Ironworkers)

		UNIT	LABOR	MATERIAL
.1	BEAMS AND COLUMNS			
.11	Wide Flange Shapes, H-Bearing Piles		(Subcontracted)	
	Light Weight Flange Columns & Misc. Columns			
	Light Beams - American Standard			
	Small Job (5 to 30 ton)	Ton	488.00	1,624.00
	Medium Job (30 to 100 ton)	Ton	477.00	1,590.00
	Large Job (100 ton and up)	Ton	445.00	1,546.00
.12	Junior Beams	Ton	461.00	1,658.00
.2	CHANNELS - American Std. - Junior -3" to 15"	Ton	625.00	1,456.00
	10" to 12"	Ton	604.00	1,635.00
.3	ANGLES (Equal and Unequal)	Ton	678.00	1,792.00
.4	STRUCTURAL T's (from wide Flange Shapes)			
	Light Shapes and American Standard	Ton	488.00	1,568.00
.5	STRUCTURAL TUBING (Steel Pipe, Square & Rectangular)	Ton	498.00	1,600.00
.6	PLATES AND BARS	Ton	519.00	1,680.00
.7	BOX GIRDERS	Ton	837.00	1,613.00
.8	TRUSSES	Ton	880.00	2,184.00
	Add for Welded Field Connections	Ton	196.00	-
	Add for High Strength Bolting	Ton	-	207.00
	Add for Riveted	Ton	371.00	-
	Add for Weathering Steel	Ton	-	291.00
	Add for Galvanizing	Ton	-	420.00

0502.0 MISCELLANEOUS METALS (L+M) (Ironworkers)

		UNIT	LABOR	MATERIAL
.1	BOLTS (with Nuts & Washers in Concrete or Masonry) See 0201.36 for Hole Drilling in Field			
	1/2" x 4"	Each	2.33	1.90
	5/8" x 4"	Each	2.44	2.10
	3/4" x 4"	Each	2.65	2.70
	1" x 4"	Each	2.86	3.50
.2	CASTINGS			
	Manhole Ring & Cover -24" Light Duty	Each	87.00	302.00
	24" Heavy Duty	Each	98.00	367.00
	Wheel Guards- 2' 6"	Each	91.00	232.00
.3	CORNER GUARDS- 3" x 3" x 1/4" Steel	LnFt	5.70	7.20
	4" x 4" x 1/4" Steel	LnFt	6.40	9.20
	Galv.	LnFt	5.80	11.20
.4	FIRE ESCAPES- 24" Steel Grate	Per Floor	1,092.00	3,240.00
		Tread	68.00	194.00
.5	FRAMES, CHANNEL- 6" - 8.2 lbs	LnFt	4.90	12.20
	8" - 11.5 lbs	LnFt	6.80	16.60
	10" - 15.3 lbs	LnFt	9.30	29.00
.6	GRATINGS- Steel - 1" x 1/8"	SqFt	2.40	8.60
	Aluminum - 1" x 1.8"	SqFt	2.40	17.80
	Cast Iron #2 - Light Duty	SqFt	2.40	12.00
	Heavy Duty	SqFt	2.10	23.00
.7	GRILLES	SqFt	3.70	10.30
.8	HANDRAILS - 1 1/2"			
	Wall Mounted - Single Rail- Steel	LnFt	6.78	12.00
	Aluminum	LnFt	7.95	44.00
	Floor Mounted - 2 Rail- Steel	LnFt	9.20	26.50
	3 Rail - Steel	LnFt	10.20	32.60
	Add for Galvanizing	LnFt	-	20%
.9	LADDERS - Steel	LnFt	12.45	41.00
	Aluminum	LnFt	12.00	5.40
.10	LINTELS, LOOSE	Ton	572.00	1,254.00
	3" x 3" x 1/4" - 4.9 lbs per foot	LnFt	1.90	4.70
	4" x 3" x 1/4" - 5.8 lbs per foot	LnFt	2.33	5.30
	4" x 3 1/2" x 1/4" - 6.2 lbs per foot	LnFt	2.50	5.60
	4" x X3 1/2" x 5/16" - 7.7 lbs per foot	LnFt	3.20	6.50
	4" x X3 1/2" x 3/8" - 9.1 lbs per foot	LnFt	3.65	7.60
	4" x 4" x 1/4" - 6.6 lbs per foot	LnFt	3.00	6.30
	Add for Punching Holes	Each	-	4.40
	Add for Galvanizing	LnFt	-	1.30

		UNIT	LABOR	MATERIAL (Subcontracted)
0502.0	**MISCELLANEOUS METALS (Cont'd...)**			
.11	PIPE GUARDS - 4" - 4' Above Grade &	LnFt	10.65	9.30
	(Bollards in Concrete) - 4' Below Grade	LnFt	11.20	12.00
.12	PLATE, CHECKERED - 1/4" - 10.2 lbs	SqFt	1.90	9.80
	1/8" - 6.2 lbs	SqFt	1.50	8.80
.13	STAIRS, METAL PAN - 4' Wide x 10' (No Concrete)	PerFl	954.00	4,428.00
	Including Supports	Tread	67.00	297.00
.14	STAIRS, CIRCULAR - 3' x 10' - Aluminum	PerFl	535.00	4,536.00
	Steel	PerFl	588.00	4,752.00
	Cast Iron	PerFl	647.00	5,152.00
.15	WIRE GUARDS	SqFt	2.50	11.70
.16	TRENCH FRAME AND COVER - 2'	LnFt	6.60	12.42
0503.0	**ORNAMENTAL METALS (M) (Ironworkers)**			
	Non Ferrous - See 0502.0			
0503.0	**OPEN WEB JOISTS (L+M) (Ironworkers)**			
.1	STANDARD BAR JOISTS (Spans to 48')			
	Small Jobs (to 20 ton) - Spans to 15'	Ton	382.00	1,232.00
	25'	Ton	387.00	1,210.00
	35'	Ton	371.00	1,177.00
	Medium Jobs (20 - 50 ton) - Spans to 15'	Ton	371.00	1,232.00
	25'	Ton	366.00	1,199.00
	35'	Ton	360.00	1,177.00
	48'	Ton	350.00	1,144.00
	Large Jobs (50 ton & up) - Spans to 15'	Ton	360.00	1,221.00
	25'	Ton	350.00	1,119.00
	35'	Ton	350.00	1,177.00
	48'	Ton	340.00	1,166.00
.2	LONG SPAN JOISTS (Spans 48' to 150')			
	Small Jobs (to 30 ton) - Spans to 60'	Ton	382.00	1,221.00
	Medium Jobs (30 -100 ton) - Spans to 60'	Ton	376.00	1,232.00
	80'	Ton	360.00	1,260.00
	100'	Ton	340.00	1,271.00
	Large Jobs (100 ton & Up) - Spans to 60'	Ton	355.00	1,221.00
	80'	Ton	350.00	1,254.00
	100'	Ton	345.00	1,287.00
	120'	Ton	334.00	1,364.00
	150'	Ton	329.00	1,419.00
0505.0	**METAL DECKING (L+M) (Sheet Metal Workers)**			
.1	SHORT SPAN (Spans to 6')			
	(Baked Enamel - Ribbed, V Beam & Seam)			
	1 1/2" Deep - 18 Ga	Sq	57.00	173.00
	20 Ga	Sq	53.00	135.00
	22 Ga	Sq	52.00	130.00
	Add for Galvanized	Sq	-	10%
.2	LONG SPAN (Spans to 12') (Enameled & Open Bottom)			
	3" Deep - 16 Ga	Sq	74.00	529.00
	18 Ga	Sq	69.00	475.00
	20 Ga	Sq	70.00	421.00
	22 Ga	Sq	67.00	346.00
	4 1/2" Deep - 16 Ga	Sq	79.60	626.00
	18 Ga	Sq	78.65	583.00
	20 Ga	Sq	77.20	529.00
	Add for Acoustical	Sq	-	20%
	Add for Galvanized	Sq	-	20%

		UNIT	LABOR	MATERIAL
			(Subcontracted)	
0505.0	**METAL DECKING (L+M) (Sheet Metal), Cont'd...**			
.3	CELLULAR - Electric and Air Flow			
	3" Deep - 16 - 16 Ga Galvanized	Sq	89.70	713.00
	18 - 18 Ga Galvanized	Sq	83.40	680.00
	4 1/2" Deep- 16 - 16 Ga Galvanized	Sq	104.40	950.00
	18 - 18 Ga Galvanized	Sq	97.70	800.00
.4	CORRUGATED METAL			
	Standard _(Spans to Approx.)_			
	Black - .015" 3' 6"	Sq	43.00	81.00
	Galvanized - .015" 3' 6"	Sq	43.00	97.00
	Heavy Duty			
	Black - 26 Ga 5' 0"	Sq	46.30	95.00
	Galvanized - 26 Ga 5' 0"	Sq	46.30	100.00
	Super Duty			
	Black - 24 Ga 7' 0"	Sq	48.55	137.00
	Galvanized - 24 Ga 7' 0"	Sq	48.55	147.00
	Black - 22 Ga 7' 6"	Sq	49.70	158.00
	Galvanized - 22 Ga 7' 6"	Sq	49.70	162.00
.5	BOX RIB			
	2" - 24 Ga	Sq	67.10	481.00
	22 Ga	Sq	71.70	508.00
	20 Ga	Sq	76.10	540.00
0506.0	**METAL SIDINGS & ROOFINGS (L+M) (Sheet Metal)**			
.1	INDUSTRIAL			
.11	Steel (24 Ga - See 0505 for Other Gauges)			
	Corrugated - Galvanized	Sq	83.30	141.00
	Ribbed - Enameled	Sq	76.50	137.00
	Add for Liner Panel	Sq	38.05	128.00
	Add for Insulation	Sq	32.50	34.00
.12	Aluminum- Corrugated	Sq	72.00	128.00
.2	ARCHITECTURAL (Prefinished)			
.21	Steel - Baked Enameled	Sq	109.90	230.00
	Porcelain Enameled	Sq	139.00	284.00
	Acrylic Enameled	Sq	133.00	246.00
	Plastic Faced	Sq	133.00	310.00
.22	Aluminum - Anodized	Sq	129.00	230.00
	Plastic Faced	Sq	131.00	310.00
	Porcelainized	Sq	136.00	300.00
	Add for Liner Panel and Insulation	Sq	51.00	134.00
.23	Stainless Steel	Sq	214.00	696.00
.24	Protected Metal (Asphalt on Steel with Finish)	Sq	157.00	482.00
	See 1310 for Metal Buildings			
0507.0	**LIGHT GAGE FRAMING (L+M) (Ironworkers)**			
	3 5/8"- 16 Ga	LnFt	.69	4.20
	6"- 16 Ga	LnFt	.72	5.20
	8"- 16 Ga	LnFt	.87	6.05
	10"- 16 Ga	LnFt	1.04	7.60
	Average - Wall	Lb	.40	1.10
	Average - Hanging	Lb	.82	1.30
0508.0	**EXPANSION CONTROL (L+M) (Ironworkers)**			
	4"- Aluminum	LnFt	8.05	43.75
	Bronze	LnFt	8.50	121.00
	Stainless Steel	LnFt	8.50	124.00
	2"- Aluminum	LnFt	6.50	39.40
	Bronze	LnFt	8.00	71.70
	Stainless Steel	LnFt	8.00	79.40

		Sq. Ft. Cost
0501.0	**Structural Steel Frame**	
	To 30 Ton 20' Span	8.10
	24' Span	8.60
	28' Span	9.10
	32' Span	10.30
	36' Span	11.44
	40' Span	12.40
	44' Span	13.10
	48' Span	14.10
	52' Span	15.90
	56' Span	17.50
	60' Span	18.50
	Deduct for Over 30 Ton	5%
0504.0	**Open Web Joists**	
	To 20 Ton 20' Span	2.64
	24' Span	2.70
	28' Span	2.75
	32' Span	2.90
	36' Span	3.10
	40' Span	3.63
	48' Span	4.51
	52' Span	5.83
	56' Span	5.39
	60' Span	5.94
	Deduct for Over 20 Ton	10%
0505.0	**Metal Decking**	
	1 1/2" Deep - Ribbed - Baked Enamel - 18 Ga	3.24
	20 Ga	3.13
	22 Ga	2.70
	3" Deep - Ribbed - Baked Enamel - 18 Ga	6.53
	20 Ga	6.20
	22 Ga	5.60
	4 1/2" Deep - Ribbed - Baked Enamel - 16 Ga	8.50
	18 Ga	7.30
	20 Ga	6.80
	3" Deep - Cellular - 18 Ga	8.30
	16 Ga	10.40
	4 1/2" Deep - Cellular - 18 Ga	11.70
	16 Ga	13.30
	Add for Galvanized	10%
	Corrugated Black Standard .015	1.62
	Heavy Duty - 26 Ga	1.73
	S. Duty - 24 Ga	2.30
	22 Ga	2.40
	Add for Galvanized	15%
0506.0	**Metal Sidings and Roofings**	
	Aluminum- Anodized	4.35
	Porcelainized	8.75
	Corrugated	2.90
	Enamel, Baked Ribbed	4.50
	24 Ga Ribbed	3.10
	Acrylic	5.35
	Porcelain Ribbed	5.90
	Galvanized- Corrugated	2.90
	Plastic Faced	6.40
	Protected Metal	7.70
	Add for Liner Panels	2.40
	Add for Insulation	.80

0601.0 ROUGH CARPENTRY (Carpenters)

		UNIT	LABOR	MATERIAL
.1	LIGHT FRAMING & SHEATHING (Standard - Hem-Fir)			
.11	Joists, Beams and Headers - 16" O.C. to 14'			
	2"x 6" Floor	BdFt	.65	.54
	2"x 8" Floor	BdFt	.63	.57
	2"x10" Floor	BdFt	.60	.59
	2"x12" Floor	BdFt	.63	.61
	Add for Ceiling Joists - 2nd Fl or Roof	BdFt	.13	-
	Add for Door, Window & Misc. Headers	BdFt	.42	-
	Add for Sloped Installations	BdFt	40%	-
	Add for Ledgers	BdFt	.27	-
.12	Studs, Plates and Miscellaneous Framing - 16" O.C.			
	2" x 2"	BdFt	.90	.57
	2" x 3"	BdFt	.85	.55
	2" x 4"	BdFt	.73	.53
	2" x 6"	BdFt	.71	.58
	Add for 2nd Floor and Above per Floor	BdFt	.13	-
	Add for Bolted Plates or Sills	BdFt	.64	-
	Add for Each 1' Above 8' in Length	BdFt	.04	-
	Add for Fire Stops, Fillers, and Nailers	BdFt	.40	-
	Add for Soffit & Suspended Framing 2"x 4"	BdFt	.90	-
	See 0602 for Metal Studs and Plates			
.13	Bridging - 16" O.C.			
	1" x 3" Wood Diagonal	Set	1.61	.42
	2" x 4" Diagonal	Set	1.82	1.00
	1" Metal Diagonal - 18 Ga	Set	1.61	1.00
	2" x 8" Solid	BdFt	1.89	.71
.14	Rafters and Field-Made Trussed Rafters - 16" O.C.			
	(See 0605 for Prefabrication)			
	2" x 4"	BdFt	.95	.62
	2" x 6"	BdFt	.93	.60
	2" x 8"	BdFt	.90	.57
	2" x 10"	BdFt	.85	.61
	2" x 4" Trusses - Make up in Field	BdFt	.74	.78
	Erection	BdFt	.51	-
	Add for Hips and Valleys	BdFt	.51	-
	Add for Bracing	BdFt	.95	.65
.15	Stairs			
	2" x 10" and 2" x 12" for Stringers	BdFt	2.10	.61
	1" x 8" and 2" x 10" for Risers & Stairs	BdFt	1.46	.62
	Average for Stringers, Treads and Risers	BdFt	1.75	.76
	Add to Above Framing Where Applicable:			
	Circular Installations	BdFt	60%	-
	Diagonal Installations	BdFt	10%	7%
	Kiln Dry Lumber	BdFt	-	15%
	Construction Grade Lumber	BdFt	-	.06
	Lumber Over 14'	BdFt	-	.05
	Winter Construction	BdFt	10%	-
	Add for Pressure Treating			
	2" x 4" - 2" x 6" - 2" x 8"	BdFt	-	.17
	2" x 10" - 2" x 12"	BdFt	-	.26
	Add for Fire Treating - .40	BdFt	-	.17
	- Dricon	BdFt	-	.32

0601.0   ROUGH CARPENTRY, Cont'd...

		UNIT	LABOR	MATERIAL
.16	Sub Floor Sheathing (Structural)			
	1" x 8" and 1" x 10" #3 Pine	BdFt	.45	.70
	1/2" x 4' x 8' Plywood - CD Exterior	SqFt	.42	.57
	5/8" x 4' x 8' Plywood - CD Exterior	SqFt	.43	.67
	3/4" x 4' x 8' Plywood - CD Exterior	SqFt	.45	.77
	3/4" x 4' x 8' Plywood - T&G, CD Exterior	SqFt	.51	1.00
	1/2" x 4' x 8' OSB Board	SqFt	.45	.75
	5/8" x 4' x 8' OSB Board	SqFt	.46	.90
	3/4" x 4' x 8' OSB Board, T&G	SqFt	.50	1.10
.17	Floor Sheathing (Over Sub Floor) With Partitions in Place			
	1/4" x 4' x 8' Plywood - CD	SqFt	.46	.43
	3/8" x 4' x 8' Plywood - CD	SqFt	.47	.44
	1/2" x 4' x 8' Plywood - CD	SqFt	.48	.57
	5/8" x 4' x 8' Plywood - CD	SqFt	.49	.67
	3/8" x 4' x 8' Particle Board	SqFt	.47	.26
	1/2" x 4' x 8' Particle Board	SqFt	.48	.31
	5/8" x 4' x 8' Particle Board	SqFt	.49	.39
	3/4" x 4' x 8' Particle Board	SqFt	.50	.45
.18	Wall Sheathing			
	1" x 8" and 1" x 10" #3 Pine	BdFt	.50	.70
	3/8" x 4' x 8' Plywood - CD Exterior	SqFt	.49	.46
	1/2" x 4' x 8' Plywood - CD Exterior	SqFt	.50	.57
	5/8" x 4' x 8' Plywood - CD Exterior	SqFt	.51	.67
	3/4" x 4' x 8' Plywood - CD Exterior	SqFt	.53	.78
	1/2" x 4' x 8' Fiberboard Impregnated	SqFt	.48	.27
	25/32" x 4' x 8' Fiberboard Impregnated	SqFt	.50	.30
	1" x 2' x 8' T&G Styrofoam	SqFt	.51	.34
	2" x 2' x 8' T&G Styrofoam	SqFt	.68	.65
	1/2" x 4' x 8' OSB Board	SqFt	.49	.75
	5/8" x 4' x 8' OSB Board	SqFt	.52	.90
	3/8" x 4' x 8			1.10
.19	Roof Sheathing - Flat			
	1" x 6" and 1" x 8" #3 Pine	BdFt	.50	.70
	1/2" x 4' x 8' Plywood - CD Exterior	SqFt	.47	.57
	5/8" x 4' x 8' Plywood - CD Exterior	SqFt	.49	.67
	3/4" x 4' x 8' Plywood - CD Exterior	SqFt	.50	.77
	Add for Sloped Installation	SqFt	.15	-
	Add Steep Sloped Installation (over 5-12)	SqFt	.32	-
	Add for Dormer Work	SqFt	.64	-
	Add to Above Sheathing where Applicable:			
	Fire Treating	SqFt	-	.32
	Pressure Treating	SqFt	-	.22
	Winter Construction	SqFt	10%	-
	Plywood - AC Exterior (Good One Side)	SqFt	-	.28
	Plywood - AD Exterior (Good One Side)	SqFt	-	.34
	Diagonal Installations	SqFt	8%	7%
	Plywood 10' Lengths	SqFt	-	.23

Recommended 10% Decrease on Labor Units for Residential Construction

0601.0   ROUGH CARPENTRY, Cont'd...

  .2   HEAVY FRAMING AND DECKING (Construction-Grade Fir, Air Dried S4S)

      See 0604.0 for Laminated Heavy Framing & Decking.

      See 0606.0 for Wood Treatments.

      See 0109.8 for Wood Curbs and Walls.

	UNIT	LABOR	MATERIAL
.21 Columns and Posts			
4"x 4"	MBF	615.00	893.00
4"x 6"	MBF	610.00	914.00
6"x 6"	MBF	562.00	1,082.00
8"x 8"	MBF	557.00	1,208.00
12"x12"	MBF	530.00	1,680.00
Add for Fitting at Base and Heat	MBF	180.00	-
Beams and Joists			
4" x 4"	MBF	615.00	893.00
4" x 6"	MBF	610.00	914.00
4" x 8"	MBF	604.00	1,040.00
4" x10"	MBF	567.00	1,103.00
6" x 6"	MBF	572.00	1,092.00
6" x 8"	MBF	557.00	1,134.00
6" x10"	MBF	551.00	1,197.00
8" x12"	MBF	541.00	1,407.00
Add to 0601.21 Above for:			
Rough Sawn	MBF	-	137.00
Mill Cutting to Length and Shape	MBF	-	137.00
Nails	MBF	-	68.00
Hauling & Unloading (if Quoted Cars)	MBF	-	89.00
Equipment & Operator Cost - Fork Lift	MBF	-	74.00
Equipment & Operator Cost - Crane	MBF	-	105.00
Add for Red Cedar	MBF	-	1,208.00
Add for Redwood	MBF	-	1,428.00
.22 DECKING			
Tongue & Groove			
Fir, Hemlock & Spruce, Construction Grade (SPF)			
4" x 4"	MBF	392.00	977.00
2" x 6" and 2" x 8"	MBF	366.00	735.00
3" x 6"	MBF	382.00	893.00
4" x 6"	MBF	392.00	1,008.00
Cedar - #3 and better			
3" x 4"	MBF	376.00	1,617.00
4" x 4" - 4" x 6" - 4" x 8" and 4" x 10"	MBF	371.00	1,764.00
White Pine			
3" x 6" and 3" x 8"	MBF	376.00	1,680.00
4" x 6" and 4" x 8"	MBF	387.00	1,817.00
Add for D Grade	MBF	-	1,155.00
2" x 6" Panelized Decking - 1 1/2" x 20"			
Premium Grade	SqFt	.81	2.36
Architectural Grade	SqFt	.81	2.00
Add for Select Grade	MBF	-	5%
Add for Hemlock	MBF	-	50.00
Add for Hip Roof	MBF	74.00	-
Add for Specified Lengths	MBF	-	35.00
Add for End Matched	MBF	-	44.00
Add for Hauling and Unloading	MBF	-	85.00
Add for Nails	MBF	-	60.00

      See 0604 for Laminated Decking

      See 0604.3 for Square Foot Deck Costs

0601.0 ROUGH CARPENTRY, Cont'd...

.3 MISCELLANEOUS CARPENTRY

.31 Blocking and Bucks (2" x 4" or 2" x 6")	UNIT	LABOR	MATERIAL
Doors and Window Bucks			
Erected Before Masonry	BdFt	1.54	.55
	or, Each	26.50	9.25
Bolted to Masonry or Steel	BdFt	2.12	.55
	or, Each	36.00	9.25
Nail Driven to Steel Studs or Masonry	BdFt	1.38	.55
	or, Each	28.32	9.25
Roof Blocking - Pressure Treated			
Edges - Bolted to Concrete	BdFt	1.61	.85
Nailed to Wood	BdFt	.81	.80
Openings - Nailed	BdFt	1.80	.80
Add for 2" x 8" or 2" x 10"	BdFt	-	.05
Add to 0601.31 for Bolts - 4' O.C.	BdFt	-	.32
.32 Grounds, Furring, Sleepers & Nailers #3 Pine			
1" x 8" Fastened to Wood	LnFt	.58	.50
1" x 6" Fastened to Wood	LnFt	.55	.36
1" x 4" Fastened to Wood	LnFt	.52	.23
1" x 3" Fastened to Wood	LnFt	.49	.20
1" x 3" Fastened to Concrete/ Masonry			
Nailed	LnFt	.66	.32
Gun Driven (Incl Fasteners)	LnFt	.58	.37
Pre Clipped	LnFt	.85	.34
Add for Ceiling Work	LnFt	.19	-
2" x 4" Not Suspended Framing	BdFt	1.60	.55
Suspended Framing	BdFt	1.96	.55
2" x 2" Not Suspended Framing	BdFt	2.00	.59
Suspended Framing	BdFt	2.90	.59
.33 Cant Strips			
Pressure Treated			
Cut from 4" x 4" - At Roof Edges	BdFt	.50	.45
	or, LnFt	.67	.67
4" x 4" - At Roof Openings	BdFt	.72	.45
	or, LnFt	.96	.67
6" x 6" - At Roof Edges	BdFt	1.75	2.30
	or, LnFt	1.20	1.50
6" x 6" - At Roof Openings	BdFt	2.42	2.30
	or, LnFt	1.61	1.50
Add for Bolting (Bolts - 4' O.C.)	BdFt	.72	.58
Not Treated			
Cut from 4" x 4" - At Roof Edges	BdFt	.49	.37
	or, LnFt	.66	.55
6" x 6" - At Roof Edges	BdFt	1.75	1.75
	or, LnFt	1.20	1.15
.34 Building Papers and Sealers			
15" Felt (432 Sq.Ft. to Roll)	SqFt	.07	.03
20" Felt (432 Sq.Ft. to Roll)	SqFt	.08	.04
30" Felt (216 Sq.Ft. to Roll)	SqFt	.10	.06
Cotton and Glass Asphalt Saturated	SqFt	.08	.06
Polyethylene 4 mil	SqFt	.07	.02
6 mil	SqFt	.08	.03
30# Red Rosin Paper	SqFt	.08	.14
Sill Sealer 6" x 100'	SqFt	.27	.22

0602.0 FINISH CARPENTRY (Carpenters)

.1   FINISH SIDINGS AND FACING MATERIAL (Exterior)
(See Quick Estimating for Square Foot Costs)

	UNIT	LABOR	MATERIAL
.11 Boards, Beveled and Lap Siding			
Cedar - Beveled - Clear 1/2" x 4"	BdFt	1.20	1.70
1/2" x 6"	BdFt	1.09	1.80
Beveled - Rough Sawn 7/8" x 8"	BdFt	1.00	1.55
7/8" x 10"	BdFt	.96	1.80
7/8" x 12"	BdFt	.94	1.85
Board- Rough Sawn 3/4" x 12"	BdFt	.92	1.80
Redwood Beveled- Clear Heart			
1/2" x 6"	BdFt	1.08	2.05
1/2" x 8"	BdFt	1.00	2.15
5/8" x 10"	BdFt	.95	2.90
3/4" x 6"	BdFt	.99	2.70
3/4" x 8"	BdFt	.94	2.80
Board - Clear Heart 3/4" x 6"	BdFt	.91	3.20
3/4" x 8"	BdFt	.88	4.40
Board, - T&G 1" x 4" & 6"	BdFt	.94	3.70
Rustic Beveled - 1" x 8" & 10"	BdFt	.91	3.90
5/4"x 6" & 8"	BdFt	.88	3.80
Add for Metal Corners	Each	.95	.40
Add for Mitering Corners	Each	3.10	-
.12 Plywood - Siding			
Fir - AC - Smooth One Side 1/4"	SqFt	.61	.81
3/8"	SqFt	.64	.92
Fir - Rough Faced 3/8"	SqFt	.66	.98
Fir - AC - Smooth One Side 1/2"	SqFt	.67	.91
5/8"	SqFt	.70	1.01
3/4"	SqFt	.74	1.11
Fir - Grooved and Rough Faced 5/8"	SqFt	.72	1.10
Cedar - Rough Sawn 3/8"	SqFt	.65	1.10
5/8"	SqFt	.72	1.35
3/4"	SqFt	.75	1.52
Cedar - Grooved & Rough Faced (4" & 8") 5/8"	SqFt	.72	1.55
Add for 4' x 10' and 4' x 9' Sheets	SqFt	-	.15
Add for Wood Batten Strips 1" x 2" Pine	LnFt	.64	.30
Add for Horizontal Joint Flashing (Spline)	LnFt	.51	.40
Add for Medium Density Fir	SqFt	.02	.16
.13 Hardboard - Paneling and Lap Siding (Primed)			
7/16" - Rough Textured 4' x 8' Paneling	SqFt	.55	1.03
7/16" - Grooved 4' x 8' Paneling	SqFt	.61	1.03
7/16" - Stucco Board 4' x 8' Paneling	SqFt	.63	1.10
Lap Siding - 7/16" x 8"	BdFt	.84	.74
7/16"x 12"	BdFt	.91	.79
Add for Prefinishing	BdFt	-	.17
.14 Shingles - Wood			
16" Red Cedar - 12" to Weather - #1	SqFt	.87	2.00
#2	SqFt	.84	1.58
#3	SqFt	.75	1.21
Fancy Butt Red Cedar 7" #1	SqFt	.89	4.10
Red Cedar Hand Split - #1 1/2" to 3/4"	SqFt	1.04	1.58
3/4" to 1 1/4" - #1	SqFt	1.08	1.90
3/4" to 1 1/4" - #2	SqFt	1.05	2.00
3/4" to 1 1/4" - #3	SqFt	1.02	1.90
Add for Fire Retardant	SqFt	-	.44
Add for 3/8" Backer Board	SqFt	.32	.25
Add for Metal Corners	Each	.87	.37
Add for Ridges, Hips and Corners	LnFt	1.46	.90
Add for 8" to Weather	SqFt	25%	-

See Division 0708 for Roof Shingles
No Finish Included in Above Costs

0602.0 FINISH CARPENTRY, Cont'd...

			MATERIAL		
			Pine	Cedar	Plywood
.15 Facias and Trim	UNIT	LABOR	#2	#3	5/8" ACX
All Sizes (average)	BdFt	.85	.82	1.71	.90
1" x 3"	LnFt	.53	.20	.43	.22
1" x 4"	LnFt	.61	.23	.53	.30
1" x 6"	LnFt	.72	.36	.91	.45
1" x 8"	LnFt	.77	.49	1.26	.60
1" x 12"	LnFt	.85	.81	1.71	.90
Add for Clear Grades			400%	100%	-	30%

.16 Soffits	UNIT	LABOR	MATERIAL
1/2" Plywood Fir ACV	SqFt	1.06	.65
Aluminum	SqFt	1.06	1.35
Plastic - Egg Crate	SqFt	1.06	1.40

.2   FINISH WALLS AND CEILING MATERIALS (Interior)

.21 Boards (See Quick Est. - Size/Cutting Allowance Added)			
Cedar - #3 Grade - 1" x 6"	SqFt	.88	2.00
Knotty 1" x 6"	SqFt	.86	2.10
1" x 8"	SqFt	.83	2.30
D Grade 1" x 4"	SqFt	.88	2.95
1" x 6"	SqFt	.86	3.05
1" x 8"	SqFt	.84	2.95
Aromatic 1" x 6"	SqFt	.88	3.05
Redwood - Constr.1" x 6"	SqFt	.92	2.25
1" x 8"	SqFt	.86	2.50
Add for Clear	SqFt	10%	100%
Fir - Beaded 5/8" x 4"	SqFt	.92	2.05
#2 Grade 1" x 6"	SqFt	.87	2.25
1" x 8"	SqFt	.85	2.20
1" x 10"	SqFt	.78	2.25

.22 Hardboard & Pressed Woods (4' x 8' Panels)			
Tempered 1/8"	SqFt	.56	.32
1/4"	SqFt	.63	.41
Pegboard (Perforated)1/8"	SqFt	.58	.38
1/4"	SqFt	.64	.51
Plastic Faced 1/4"	SqFt	.67	.63
Flake Board 1/8"	SqFt	.52	.32
Add for Metal or Plastic Mouldings	LnFt	.54	.42
Add for Pre-finished	SqFt	-	.19

.23 Plywood Paneling (4'x 8' Panels, Prefinished)			
Birch, Natural 1/4"	SqFt	.90	1.05
3/4"	SqFt	1.02	1.75
White Select 1/4"	SqFt	.91	1.70
Oak, Red - Rotary Cut 1/4"	SqFt	.91	.95
3/4"	SqFt	1.04	2.10
White 1/4"	SqFt	.90	2.10
Cedar, Aromatic 1/4"	SqFt	.91	1.75
Cherry 3/4"	SqFt	1.04	2.40
Pecan 3/4"	SqFt	1.04	2.40
Walnut 1/4"	SqFt	.90	2.40
3/4"	SqFt	1.04	3.80
Mahogany, Lauan 1/4"	SqFt	.90	.75
3/4"	SqFt	1.04	1.30
African 3/4"	SqFt	1.04	2.60
Knotty Pine 1/4"	SqFt	.90	1.20
3/4"	SqFt	1.04	2.00
Slat Wall - Painted 3/4"	SqFt	1.04	1.60
Oak 3/4"	SqFt	1.10	2.50
Add for V Groove and T&G	SqFt	-	.28
Add for Glued-On Applications	SqFt	.18	.11
Add for 10' Lengths	SqFt	.13	.26

0602.0 FINISH CARPENTRY, Cont'd...

.24 Gypsum Board and Framing
(See 0902 - Subcontracted Work Costs)
(See 0902 - Metal Framing SqFt Costs)

	UNIT	LABOR	MATERIAL
Finish Board (Walls to 8' - Screwed On)			
3/8" 4' x 8' and 12'	SqFt	.36	.24
1/2" 4' x 8' and 12'	SqFt	.37	.26
5/8" 4' x 8' and 12	SqFt	.38	.30
1" 4' x 8' and 12' Core Board	SqFt	.39	.55
1" 4' x 8' and 12' Bead Board	SqFt	.37	.55
Add for Adhesive Application	SqFt	.08	.02
Deduct for Nail-on Application	SqFt	.02	-
Add for: Fire Resistant Board	SqFt	.02	.08
Add for: Foil Backed (Insulating)	SqFt	.02	.11
Add for: Moisture Resistant Type 5/8" & 1/2"	SqFt	.02	.07
Add for: Work Above 8'	SqFt	.06	-
Add for: Work Over 2-Story per Story	SqFt	.06	-
Add for: Ceiling Work - To Wood	SqFt	.15	-
To Steel	SqFt	.36	-
Add for: Resilient Clip Application	SqFt	.35	.08
Add for: Small Cut Up Areas	SqFt	.38	.06
Add for: Beam and Column Work	SqFt	.49	.06
Add for: Circular Construction	SqFt	.42	.08
Add for: Vinyl Faced Board			
3/8" x 4' x 8' - Cost Variable	SqFt	.42	.83
1/2" x 4' x 8' - Cost Variable	SqFt	.47	.90
Metal Framing (Studs, Runners and Channels)			
Studs 25-Gauge 1 5/8"	LnFt	.53	.32
2 1/2"	LnFt	.54	.35
3 5/8"	LnFt	.61	.38
6"	LnFt	.56	.51
8"	LnFt	.61	.54
Studs 209A 1 5/8"	LnFt	.51	.43
2 1/2"	LnFt	.53	.49
3 5/8"	LnFt	.55	.55
6"	LnFt	.56	.73
8"	LnFt	.60	.81
7/8" Furring Channels	LnFt	.53	.35
1 1/2" Furring Channels (Cold Rolled)	LnFt	.54	.49
Resilient Channels	LnFt	.51	.38
Add for Over 10' Lengths	LnFt	.18	.09
Add for 20-Gauge 6" Material	LnFt	.17	.49
Add for 16-Gauge 8" Material	LnFt	.30	1.08
Metal Trim			
Casing Bead	LnFt	.72	.24
Corner Bead	LnFt	.55	.15
L & J Bead 3/8" - 1/2" - 5/8"	LnFt	.55	.23
Expansion Bead - Metal	LnFt	.55	.65
Plastic	LnFt	.55	.32
Metal Mouldings	LnFt	.85	.56
Taping and Sanding	SqFt	.32	.07
Texturing	SqFt	.40	.08
Thin Coat Plaster – or Skin Coat	SqFt	.90	.29
Sound Deadening Insulation - 2 1/2"	SqFt	.32	.27
3 1/2"	SqFt	.37	.27
6"	SqFt	.39	.35
.25 Cement Board 1/2" 3' x 5 1/2'	SqFt	.37	1.15
1/2" 4' x 8'	SqFt	.37	1.15

0603.0 MILLWORK & CUSTOM WOODWORK (M) (Carpenters)   MATERIAL

.1 CUSTOM CABINET WORK

			Prefinished Red Oak or Birch	Plastic Laminate
See Division 11 - Stock Cabinets (Medium etc)				
Base Cabinets 35" High x 24" Deep -	UNIT	LABOR		
18" W	LnFt	18.40	114.00	109.00
24" W	LnFt	18.40	109.00	104.00
30" W	LnFt	17.40	105.00	100.00
36" W	LnFt	17.40	100.00	95.00
36" Wide Sink Fronts	LnFt	16.20	71.00	66.00
42"	LnFt	16.20	77.00	71.00
36" Wide Corner Cabinet	LnFt	17.30	126.00	121.00
Add for Lazy Susan	Each	23.10	116.00	110.00
Add per Drawer	Each	3.70	24.00	23.00
Upper Cabinets - 12" Deep				
30" High 18" Wide	LnFt	18.40	83.00	80.00
24" W	LnFt	18.40	78.00	75.00
30" W	LnFt	17.40	76.00	72.00
36" W	LnFt	17.40	72.00	69.00
24" High 18" Wide	LnFt	17.40	71.00	68.00
24" W	LnFt	17.40	66.00	63.00
30" W	LnFt	17.40	64.00	61.00
36" W	LnFt	16.20	63.00	60.00
15" High 18" Wide	LnFt	16.20	65.00	62.00
24" W	LnFt	16.20	62.00	59.00
30" W	LnFt	14.95	61.00	58.00
36" W	LnFt	14.95	60.00	57.00
Utility Cabinets - 84" H x 24" D				
18" Wide	LnFt	22.00	168.00	145.00
24" W	LnFt	22.00	158.00	142.00
Add - Prefinished Interior	-	-	8%	-
Deduct - Prefinished to Unfinished	-	-	12%	-
Add for Plastic Lam. Interior	-	-	-	100%
Add to Above for Special Hdwe.	Unit	8.06	-	24.00
Add - White/Birch/White Oak/Maple	-	-	-	20%
Add - Walnut	-	-	-	100%
China or Corner Cabinet 84"H x 36"W	Each	117.00	656.00	650.00
Oven Cabinet 84" High x 27" High	Each	111.00	515.00	500.00
Vanity Cabinets 30" High x 21" Deep				
30" Wide	LnFt	34.70	110.00	107.00
36" W	LnFt	35.80	107.00	104.00
48" W	LnFt	39.00	103.00	103.00

.2 COUNTER TOPS, 25" with Back Splash 4"   MATERIAL

	UNIT	LABOR	MATERIAL
Plastic Laminated 1 1/2"	LnFt	11.00	30.00
Deduct for No Back Splash	LnFt	2.40	5.25
Plastic Laminated Vanity Top	LnFt	9.90	34.00
Marble	LnFt	10.40	65.00
Artificial	LnFt	10.40	58.00
Wood Cutting Block, 1 1/2"	LnFt	10.50	67.00
Stainless Steel	LnFt	12.20	107.00
Polyester-Acrylic Solid Surface	LnFt	11.00	126.00
Granite - Artificial - 1 1/4"	LnFt	13.40	68.00
3/4"	LnFt	12.20	51.00
Quartz	LnFt	12.20	63.00
Plastic (Polymer)	LnFt	12.00	42.00

0603.0 MILLWORK & CUSTOM WOODWORK, Cont'd...
.3 CUSTOM DOOR FRAMES & TRIM

	Unit	Labor	Birch	Fir	Poplar	Oak	Pine	Maple
Custom Jamb & Stop								
4 5/8" x 2'6" x 3/4" x 6'8"	Each	$42	$124	$65	$64	$116	$116	-
x 2'8" x 3/4"	Each	44	126	67	65	118	116	-
x 3'0" x 3/4"	Each	45	122	72	66	119	121	-
5 1/4" x 2'6" x 3/4" x 6'8"	Each	44	134	77	66	123	118	-
x 2'8" x 3/4"	Each	45	137	79	68	126	119	-
x 3'0" x 3/4"	Each	46	141	82	71	142	126	-
6 3/4" x 2'6" x 3/4" x 6'8"	Each	46	184	-	-	185	140	-
x 2'8" x 3/4"	Each	47	189	-	-	187	141	-
x 3'0" x 3/4"	Each	48	191	-	-	189	144	-
Stock Jamb & Stop								
4 5/8" x 2'6" x 3/4" x 6'8"	Each	40	110	38	-	108	69	-
x 2'8" x 3/4"	Each	41	112	39	-	110	72	-
x 3'0" x 3/4"	Each	40	114	40	-	113	74	-
5 1/4" x 2'6" x 3/4" x 6'8"	Each	41	121	44	52	121	75	-
x 2'8" x 3/4"	Each	42	123	45	54	123	76	-
x 3'0" x 3/4"	Each	44	128	46	55	133	78	-
6 3/4" x 2'6" x 3/4" x 6'8"	Each	44	151	-	-	168	105	-
x 2'8" x 3/4"	Each	45	153	-	-	170	107	-
x 3'0" x 3/4"	Each	46	160	-	-	172	110	-
Fire Rated Jamb & Stop								
4 5/8" x 3'0" x 1 1/16"	Each	45	194	-	-	179	-	-
5 1/4" x 3'0" x 1 1/16"	Each	46	203	-	-	190	-	-
6 3/4" x 3'0" x 1 1/16"	Each	48	208	-	-	195	-	-
Trim Only -11/16" x 2 1/4"	Each	41	118	32	42	102	83	-
(2 sides) - 3/4" x 3 1/2"	Each	48	189	38	59	133	108	-
.4 MOULDINGS & TRIM-STOCK								
Apron 7/16" x 2"	LnFt	.89	1.71	-	.49	1.05	.72	-
11/16" x 2 1/2"	LnFt	1.07	3.81	-	-	1.63	1.33	1.45
Astragal 1 3/4" x 2 1/4"	LnFt	1.12	-	-	-	5.67	4.65	-
Base 7/16" x 2 3/4"	LnFt	.89	3.81	.68	.80	1.39	1.21	1.46
9/16" x 3 1/4"	LnFt	.95	3.25	-	1.08	2.49	1.44	1.58
Base Shoe 7/16" x 2 3/4"	LnFt	.90	-	-	.47	.80	.50	.81
Batten Strip 5/8" x 1 5/8"	LnFt	.80	-	-	-	1.22	1.00	-
Brick Mould 1 1/4" x 3"	LnFt	1.18	-	1.10	-	-	1.58	-
Casing 11/16" x 2 1/4"	LnFt	.95	1.51	.82	.73	1.15	1.15	1.11
3/4" x 3 1/2"	LnFt	1.02	2.68	.92	.79	2.23	2.03	-
Chair Rail 5/8" x 1 3/4"	LnFt	.82	1.94	-	1.24	1.84	1.50	-
or Dado 11/16" x 2 1/4"	LnFt	.87	2.61	-	-	2.29	1.51	-
Closet Rod 1 5/16"	LnFt	1.31	-	.92	-	1.37	-	-
Corner Bead 3/4" x 3/4"	LnFt	.79	-	-	.67	1.19	.91	1.17
1 1/8" x 1 1/8"	LnFt	.86	-	-	1.18	1.73	1.58	1.64
Cove Mould 3/4" x 7/8"	LnFt	1.23	1.28	-	.57	.90	.72	1.01
Crown & Cove 9/16" x 2 3/4"	LnFt	1.33	3.07	-	1.12	1.95	1.74	2.14
9/16" x 3 5/8"	LnFt	1.42	3.78	-	-	3.22	2.29	-
11/16" x 4 5/8"	LnFt	1.42	-	-	-	3.78	3.50	-
Drip Cap 11/16" x 1 3/4"	LnFt	.98	-	-	-	-	1.55	-
Half Round 1/2" x 1"	LnFt	1.11	-	-	-	-	1.11	-
Hand Rail 1 5/8" x 1 3/4"	LnFt	1.55	-	1.49	-	-	1.66	-
Hook Strip 5/8" x 2 1/2"	LnFt	.92	-	.92	.97	-	1.09	-
Picture Mould 3/4" x 1 1/2"	LnFt	.92	.97	-	-	1.10	1.09	-
Quarter Round 3/4" x 3/4"	LnFt	.79	-	.42	.57	.90	.58	.91
1/2" x 1/2"	LnFt	.74	-	.33	-	-	.38	-
Sill Casement 3/4" x 1 7/8"	LnFt	1.26	-	-	-	-	1.55	-
Stool 11/16" x 2 1/2"	LnFt	1.26	3.57	1.18	1.44	3.09	1.96	-

Above prices based on 200 LnFt. Deduct for Over 200 LnFt - 20%.
Add for Walnut Trim - 100% Add for Custom Trim – 50%

0603.0   MILLWORK & CUSTOM WOODWORK, Cont'd...

		Unit	Labor	Birch		Oak	Pine	Walnut
.5	STAIRS							
	Treads 1 1/6" x 10 1/4"	LnFt	1.96			4.70	3.70	-
	Risers 3/4" x 7 1/2"	LnFt	1.75			5.15	2.60	-
	Skirt Bds 9/16" x 9 1/2"	LnFt	2.25			6.60	2.70	-
	Nosings 1/8" x 3 1/2"	LnFt	1.60			3.30	1.90	-
.6	SHELVING							
	12" Deep	SqFt	4.90	33.30	-	- 25.42	12.80	76.00
	8"	SqFt	5.00	27.95	-	- 23.25	11.70	62.00
.7	CUSTOM PANELING	SqFt	3.00	19.65	-	- 14.06	-	23.00
.8	THRESHOLDS 3/4" x 3 1/2"							
	Standard	LnFt	7.30	-	-	- 5.80	-	-
	With Vinyl	LnFt	7.30	-	-	- 6.30	-	-

0604.0   GLUE LAMINATED (M) (Carpenters) (F.O.B. Cars)

		UNIT	LABOR	MATERIAL
.1	ARCHES AND RIGID FRAMES: Arches	MBF	650.00	1,890.00
	Rigid Frames	MBF	660.00	2,058.00
.2	BEAMS, JOISTS & PURLINS: 4" x 6", 8"& 10"	MBF	660.00	2,384.00
	6" x 6", 8" & 10"	MBF	640.00	2,247.00
	8" x 6", 8" & 10"	MBF	630.00	2,153.00
.3	DECKS: Doug.Fir - 3" x 6" (2 1/4"x 5 1/4")	SqFt	1.90	5.25
	4" x 6" (3" x 5 1/4")	SqFt	1.96	5.67
	Cedar - 3" x 6" (2 1/4"x 5 1/4")	SqFt	2.00	6.20
	4" x 6"	SqFt	2.12	6.60
	Pine - 3" x 6" (2 1/4"x 5 1/4")	SqFt	1.90	4.50

0605.0   PREFABRICATED WOOD & PLYWOOD COMPONENTS (See 0601.22, Job Fabricated)

.1	WOOD TRUSSED RAFTERS			
	4 -12 Pitch 24" O.C. With 2' Overhang 57 lb. Loading			
	16' Span	Each	33.50	67.00
	20'	Each	35.80	69.00
	22'	Each	36.25	71.40
	24'	Each	39.00	75.05
	26'	Each	41.00	82.15
	28'	Each	45.05	97.20
	30'	Each	49.40	109.00
	32'	Each	55.65	121.30
	Add for 5-12 Pitch	Each	5%	15%
	Add for 6-12 Pitch	Each	12%	25%
	Add for Gable End Type (Flying)	Each	21.00	35.50
	Add for 16" O.C.	Each	-	10%
	Add for Hip Ends - 24' Span	Each	160.00	415.00
	Add for Energy Type	Each	-	10%
.2	WOOD FLOOR AND ROOF TRUSS JOISTS			
.21	Open Wood Web - Gusset Plate Connection			
	24" O.C. Up to 23' Span x 12" - Single Chord	LnFt	1.54	5.10
	to 24' Span x 15"	LnFt	1.60	5.20
	to 27' Span x 18"	LnFt	1.64	5.30
	to 30' Span x 21"	LnFt	1.80	5.50
	to 25' Span x 15" - Double Chord	LnFt	1.70	4.70
	to 30' Span x 18"	LnFt	2.10	4.80
	to 35' Span x 21"	LnFt	2.40	5.00
	Open Metal Web - Pin Connection			
	to 25' Span x 18"	LnFt	1.62	4.80
	to 30' Span x 22"	LnFt	1.74	4.90
	to 35' Span x 30"	LnFt	1.96	5.10
	to 40' Span x 30"	LnFt	2.44	5.25

0605.0 PREFABRICATED WOOD & PLYWOOD COMPONENTS, Cont'd...

		UNIT	LABOR	MATERIAL
.22	Plywood Web			
	24" O.C. Up to 15' Span x 9 1/2"	LnFt	1.29	2.95
	to 19' Span x 11 7/8"	LnFt	1.36	3.05
	to 21' Span x 14"	LnFt	1.52	3.40
	to 22' Span x 16"	LnFt	1.70	3.60
.3	LAMINATED VENEER STRUCTURAL BEAMS			
.31	Micro Lam (Plywood) 7 1/4" x 1 3/4"	LnFt	1.33	3.15
	9 1/2" x 1 3/4"	LnFt	1.38	4.50
	11 7/8" x 1 3/4"	LnFt	1.45	5.15
	14" x 1 3/4"	LnFt	1.54	6.20
	16" x 1 3/4"	LnFt	1.64	7.00
	18" x 1 3/4"	LnFt	1.83	7.70
.32	Glue Lam (Dimension Lumber)			
	9" x 3 1/2" Industrial	LnFt	1.38	9.55
	12" x 3 1/2"	LnFt	1.45	11.55
	15" x 3 1/2"	LnFt	1.54	13.65
	9" x 5 1/8"	LnFt	1.38	13.10
	12" x 5 1/8"	LnFt	1.57	16.30
	15" x 5 1/8"	LnFt	1.75	20.50
	18" x 5 1/8"	LnFt	1.90	25.20
	24" x 5 1/8"	LnFt	2.28	31.50
	Add for Architectural Grade	LnFt	.25	10%
	Add for 6 3/4"	LnFt	.27	5.00
	Add for Pine Treated	LnFt		4.00
.4	DECKS - 4' x 30' with Plywood 2 Sides - Insulated	SqFt	1.17	6.20

0606.0 WOOD TREATMENTS (Preservatives)

		UNIT	LABOR	MATERIAL
	SALTS - PRESSURE (Water Borne)			
	Dimension 2 x 4-2 x 6 and 2 x 8, 2 x 10-2 x 12	BdFt	-	.17
	2 x 10 and 2 x 12	BdFt	-	.27
	Plywood 5/8" CDX	SqFt	-	.24
	SALTS - NON PRESSURE	BdFt	-	.13
	CREOSOTE (Penta)	BdFt	-	.36
	FIRE RETARDANTS - PRESSURE			
	Dimension and Timbers	BdFt	-	.35
	Plywood 5/8" CDX	SqFt	-	.35

0607.0 ROUGH HARDWARE

		UNIT	LABOR	MATERIAL
.1	NAILS, BOLTS, ETC. Job Average	SqFt & BdFt	-	.02
	Common (50# Box) 8 Penny	CWT	-	55.00
	16 Penny	CWT	-	50.00
	Finish #3	CWT	-	64.00
	Add for Coated	CWT	-	60%
.2	JOIST HANGERS	Each	1.52	1.08
.3	BOLTS - 5/8" x 12"	Each	1.82	1.60

0608.0 EQUIPMENT (Saws, Woodworking, Drills, Etc.)

	Average Job - As a Percentage of Labor	-	-	3%
	Heavy Equipment in Division 2			
	See 1-6A and 1-7A for Rental Rates			

Recommend 10% Decrease on Labor Units for Residential Construction

The costs below are average, priced as total contractor/subcontractor costs with an overhead of 5% and fee of 10% included. Units include fasteners and equipment, 35% taxes and insurance on labor, and 10% General Conditions and Equipment. Also included are nails, cutting, size and lap allowances. See Example on Page 4.

0601.0 ROUGH CARPENTRY

		UNIT	COST
.1	LIGHT FRAMING AND SHEATHING		
.11	Joists and Headers - Floor Area - 16" O.C.		
	2" x 6" Joists - with Headers & Bridging	SqFt	2.15
	2" x 8" Joists	SqFt	2.70
	2" x 10" Joists	SqFt	3.05
	2" x 12" Joists	SqFt	3.78
	Add for Ceiling Joists, 2nd Floor and Above	SqFt	8%
	Add for Sloped Installation	SqFt	15%
.12	Studs, Plates and Framing - 8' Wall Height - 16" O.C.		
	2" x 3" Stud Wall - Non-Bearing (Single Top Plate)	SqFt	1.39
	2" x 4" Stud Wall - Bearing (Double Top Plate)	SqFt	1.58
	2" x 4" Stud Wall - Non-Bearing (Single Top Plate)	SqFt	1.47
	2" x 6" Stud Wall - Bearing (Double Top Plate)	SqFt	1.94
	2" x 6" Stud Wall - Non-Bearing (Single Top Plate)	SqFt	1.89
	Add for Stud Wall - 12" O.C.	SqFt	8%
	Deduct for Stud Wall - 24" O.C.	SqFt	22%
	Add for Bolted Plates or Sills	SqFt	.11
	Add for Each Foot Above 8'	SqFt	.06
	Add to Above for Fire Stops, Fillers and Nailers	SqFt	.47
	Add for Soffits and Suspended Framing	SqFt	.63
.13	Bridging - 1" x 3" Wood Diagonal	Set	2.52
	2" x 8" Solid	Each	2.73
.14	Rafters		
	2" x 4" Rafter (Incl Bracing) 3 - 12 Slope	SqFt	1.84
	4 - 12 Slope	SqFt	2.00
	5 - 12 Slope	SqFt	2.10
	6 - 12 Slope	SqFt	2.42
	Add for Hip-and-Valley Type	SqFt	.44
	Add for 1' 0" Overhang - Total Area of Roof	SqFt	.16
	2" x 6" Rafter (Incl Bracing) 3-12 Slope	SqFt	2.31
	4-12 Slope	SqFt	2.36
	5-12 Slope	SqFt	2.42
	6-12 Slope	SqFt	2.52
	Add for Hip - and - Valley Type	SqFt	.55
.15	Stairs	Each Floor	457.00
.16	Sub Floor Sheathing (Structural)		
	1" x 8" and 1" x 10" #3 Pine	SqFt	1.62
	1/2" x 4' x 8' CD Plywood - Exterior	SqFt	1.30
	5/8" x 4' x 8'	SqFt	1.46
	3/4" x 4' x 8'	SqFt	1.67
.17	Floor Sheathing (Over Sub Floor)		
	3/8" x 4' x 8' CD Plywood	SqFt	1.22
	1/2" x 4' x 8'	SqFt	1.30
	5/8" x 4' x 8'	SqFt	1.42
	1/2" x 4' x 8' Particle Board	SqFt	1.05
	5/8" x 4' x 8'	SqFt	1.14
	3/4" x 4' x 8'	SqFt	1.34

0601.0   ROUGH CARPENTRY, Cont'd...

		UNIT	COST
.18	Wall Sheathing		
	1' x 8" and 1" x 10" - #3 Pine	SqFt	1.67
	3/8" x 4' x 8' CD Plywood - Exterior	SqFt	1.27
	1/2" x 4' x 8'	SqFt	1.38
	5/8" x 4' x 8'	SqFt	1.53
	3/4" x 4' x 8'	SqFt	1.78
	25/32" x 4' x 8' Fiber Board - Impregnated	SqFt	.98
	1" x 2' x 8' T&G Styrofoam	SqFt	1.24
	2" x 2' x 8'	SqFt	1.82
.19	Roof Sheathing - Flat Construction		
	1" x 6" and 1" x 8" - #3 Pine	SqFt	1.64
	1/2" x 4' x 8' CD Plywood - Exterior	SqFt	1.41
	5/8" x 4' x 8'	SqFt	1.54
	3/4" x 4' x 8'	SqFt	1.80
	Add for Sloped Roof Construction (to 5-12 slope)	SqFt	.17
	Add for Steep Sloped Construction (over 5-12 slope)	SqFt	.30
	Add to Above Sheathing (0601.16 thru 0601.19)		
	AC or AD Plywood	SqFt	.30
	10' Length Plywood	SqFt	.22
.2	HEAVY FRAMING		
.21	Columns and Beams - 16' Span Average - Floor Area	SqFt	7.20
	20' Span	SqFt	7.50
	24' Span	SqFt	8.70
.22	Deck 2" x 6" T&G - Fir Random Construction Grade	SqFt	4.25
	3" x 6" T&G - Fir Random Construction Grade	SqFt	5.60
	4" x 6"	SqFt	7.20
	2" x 6" T&G - Red Cedar	SqFt	5.35
	Add for D Grade Cedar	SqFt	1.80
	2" x 6" T&G - Panelized Fir	SqFt	3.90
.3	MISCELLANEOUS CARPENTRY		

		EACH	or	BdFt
.31	Blocking and Bucks (2" x 4" and 2" x 6")			
	Doors & Windows - Nailed to Concrete or Masonry	50.00		2.50
	Bolted to Concrete or Masonry	56.50		3.50
	Roof Edges - Nailed to Wood	-		2.00
	Bolted to Concrete (Incl. bolts)	-		3.00

		LnFt	or	BdFt
.32	Grounds and Furring			
	1" x 6" Fastened to Wood	1.10		2.20
	1" x 4"	.95		2.90
	1" x 3"	.90		3.60
	1" x 3" Fastened to Concrete/Masonry - Nailed	1.40		5.50
	Gun Driven	1.30		5.20
	PreClipped	1.60		6.40
	2" x 2" Suspended Framing	1.40		4.30
	Not Suspended Framing	1.20		3.60

.33	Cant Strips		
	4" x 4" Treated and Nailed	1.65	-
	Treated and Bolted (Including Bolts)	2.40	-
	6" x 6" Treated and Nailed	3.30	-
	Treated and Bolted (Including Bolts)	4.20	-

		UNIT	COST
.34	Building Papers and Sealers		
	15" Felt	SqFt	.20
	Polyethylene - 4 mil	SqFt	.15
	6 mil	SqFt	.17
	Sill Sealer	SqFt	.74

0602.0 FINISH CARPENTRY

.1 FINISH SIDINGS AND FACING MATERIALS (Exterior)

.11 Boards and Beveled Sidings

	TO WEATHER	% ADDED	COST
Cedar Beveled - Clear Heart 1/2" x 4"	2 3/4"	51%	5.60
1/2" x 6"	4 3/4"	31%	4.90
1/2" x 8"	6 3/4"	23%	4.10
Rough Sawn 7/8" x 8"	6 3/4"	23%	4.25
7/8" x 10"	8 3/4"	19%	4.40
7/8" x 12"	10 3/4"	17%	4.40
3/4" x 12"	-	-	3.80
Redwood Beveled - Clear Heart			
1/2" x 6"	4 3/4"	31%	5.30
1/2" x 8"	6 3/4"	23%	5.25
5/8" x 10"	8 3/4"	19%	5.40
3/4" x 6"	4 3/4"	31%	5.50
3/4" x 8"	6 3/4"	23%	6.15
3/4" x 6" Board - Clear -		Varies	3.20
3/4" Tongue & Groove -		Varies	5.00
1" Rustic Beveled - 6" and 8"		Varies	5.60
5/4" Rustic Beveled - 6" and 8"		Varies	5.45

	UNIT	COST
Add for Metal Corners	Each	1.50
Add for Mitering Corners	Each	3.10

.12 Plywood Fir AC Smooth -One Side 1/4"

	UNIT	COST
Plywood Fir AC Smooth -One Side 1/4"	SqFt	1.90
3/8"	SqFt	2.05
1/2"	SqFt	2.10
5/8"	SqFt	2.20
3/4"	SqFt	2.40
5/8" - Grooved and Rough Faced	SqFt	2.50
Cedar - Rough Sawn 3/8"	SqFt	2.40
5/8"	SqFt	2.50
3/4"	SqFt	3.00
5/8" - Grooved and Rough Faced	SqFt	3.00
Add for Wood Batten Strips, 1" x 2" - 4' O.C.	SqFt	.40
Add for Splines	SqFt	.40

.13 Hardboard - Paneling and Lap Siding (Primed)

		UNIT	COST
3/8" Rough Textured Paneling -	4' x 8'	SqFt	2.00
7/16" Grooved -	4' x 8'	SqFt	1.90
7/16" Stucco Board Paneling		SqFt	2.10
7/16" x 8" Lap Siding		SqFt	2.40
7/16" x 12" Lap Siding		SqFt	2.30
Add for Pre-Finishing		SqFt	.18

.14 Shingles

		UNIT	COST
Wood- 16" Red Cedar - 12" to Weather	- #1	SqFt	3.90
	- #2	SqFt	3.10
	- #3	SqFt	2.90
24" - 1/2" to 3/4" Red Cedar Hand Splits	- #1	SqFt	3.33
24" - 3/4" to 1 1/4" Red Cedar Hand Splits	- #1	SqFt	3.70
	- #2	SqFt	3.33
	- #3	SqFt	3.11
Add for Fire Retardant		SqFt	.46
Add for 3/8" Backer Board		SqFt	.80
Add for Metal Corners		Each	1.46
Add for Ridges, Hips and Corners		LnFt	4.05
Add for 8" to Weather		SqFt	25%

		COST		
		LN FT	or	SQ FT
0602.0	**FINISH CARPENTRY, Cont'd...**			
.15	Facia - Pine #2 1" x 8"	1.85		2.78
	Cedar #3 1" x 8"	2.80		4.10
	Redwood - Clear 1" x 8"	4.00		6.00
	Plywood 5/8" - ACX 1" x 8"	1.90		2.90
.16	Soffit - Plywood 1/2" - ACX 1/2" Fir ACX			3.10
	Aluminum			3.10
	Plastic (Egg Crate)			3.30
.2	FINISH WALLS (Interior) (15% Added for Size & Cutting)			
.21	Boards, Cedar - #3 1" x 6" and 1" x 8"			4.06
	Knotty 1" x 6" and 1" x 8"			4.24
	D Grade 1" x 6" and 1" x 8"			5.07
	Aromatic 1" x 6" and 1" x 8"			5.25
	Redwood - Construction 1" x 6" and 1" x 8"			4.40
	Clear 1" x 6" and 1" x 8"			5.80
	Fir - Beaded 5/8" x 4"			3.90
	Pine - #2 1" x 6" and 1" x 8"			3.10
.22	Hardboard (Paneling) Tempered - 1/8"			1.25
	1/4"			1.42
.23	Plywood (Prefinished Paneling) 1/4" Birch - Natural			3.00
	3/4" Birch - Natural			3.94
	1/4" Birch - White			3.60
	1/4" Oak - Rotary Cut			2.89
	3/4"			4.10
	1/4" Oak - White			3.91
	1/4" Mahogany (Lauan)			2.10
	3/4"			3.10
	3/4" Mahogany (African)			5.20
	1/4" Walnut			4.70
.24	Gypsum Board (Paneling) - See 0902			

			COST	
			Prefinished Red Oak or Birch	Plastic Laminate
0603.0	**MILLWORK & CUSTOM WOODWORK**	UNIT		
.1	CUSTOM CABINET WORK (Red Oak or Birch)			
	Base Cabinets - Avg. 35" H x 24" D with Drawer	LnFt	188.00	184.00
	Sink Fronts	LnFt	112.00	111.00
	Corner Cabinets	LnFt	246.00	242.00
	Add per Drawer	LnFt	48.00	48.00
	Upper Cabinets - Avg. 30" High x 12" Deep	LnFt	141.00	140.00
	Utility Cabinets - Avg. 84" High x 24" Deep	LnFt	366.00	353.00
	China or Corner Cabinets - 84" High	Each	684.00	672.00
	Oven Cabinets - 84" High x 24" Deep	Each	461.00	446.00
	Vanity Cabinets - Avg. 30" High x 21" Deep	Each	244.00	237.00
	Deduct for Prefinishing Wood	Each	12%	-
.2	COUNTER TOPS - 25"			
	Plastic Laminated with 4" Back Splash	LnFt	-	55.00
	Deduct for No Back Splash	LnFt	-	10.00
	Granite - 1 1/4" - Artificial	LnFt	-	107.00
	3/4" - Artificial	LnFt	-	87.00
	Marble	LnFt	-	100.00
	Artificial	LnFt	-	91.00
	Wood Cutting Block	LnFt	-	103.00
	Stainless Steel	LnFt	-	162.00
	Polyester-Acrylic Solid Surface	LnFt	-	177.00
	Quartz	LnFt	-	95.00
	Plastic (Polymer)	LnFt	-	69.00
	(See 1103 for Stock Cabinets)			

0603.0   MILLWORK & CUSTOM WOODWORK, Cont'd...

.3   CUSTOM DOOR FRAMES - Including 2 Sides Trim

	UNIT	COST
Birch	Each	297.00
Fir	Each	183.00
Poplar	Each	189.00
Oak	Each	233.00
Pine	Each	223.00
Walnut	Each	429.00

See 0802.87 for Stock Prehung Units

.4   MOULDINGS AND TRIM

	LINEAR FOOT				
	Birch	Fir	Poplar	Oak	Pine
Apron 7/16" x 2"	3.20	-	2.09	2.74	2.18
Astragal 1 3/4" x 2 1/4"	-	-	-	6.90	5.28
Base 7/16" x 2 3/4"	3.68	1.90	2.14	2.82	3.20
9/16" x 3 1/4"	4.90	-	-	4.30	4.38
Base Shoe 7/16" x 2 3/4"	-	-	1.87	2.52	2.63
Batten Strip 5/8" x 1 5/8"	-	1.52	-	2.80	2.29
Brick Mould 1 1/14" x 2"	-	2.65	-	-	2.65
Casing 11/16" x 2 1/4"	3.40	2.10	2.30	3.20	2.89
3/4" x 3 1/2"	4.63	2.65	2.33	4.60	4.00
Chair Rail 5/8" x 1 3/4"	3.69	-	2.90	3.70	2.90
Closet Rod 1 5/16"	-	2.70	-	-	3.30
Corner Bead 1 1/8" x 1 1/8"	-	2.35	2.90	3.60	4.10
Cove Moulding 3/4" x 3/4"	3.10	-	2.50	3.00	2.50
Crown Moulding 9/16" x 3 5/8"	5.60	-	3.64	4.60	3.84
11/16" x 4 5/8"	6.49	-	-	7.20	6.10
Drip Cap	-	-	-	-	3.00
Half Round 1/2" x 1"	-	-	-	-	3.52
Hand Rail 1 5/8" x 1 3/4"	-	4.05	4.43	7.10	3.90
Hook Strip 5/8" x 2 1/2"	-	2.19	2.51	-	2.63
Picture Mould 3/4" x 1 1/2"	2.23	-	-	3.22	2.90
Quarter Round 3/4" x 3/4"	-	1.30	1.65	2.59	2.04
1/2" x 1/2"	-	1.41	-	-	1.61
Sill 3/4" x 2 1/2"	-	2.65	-	-	3.52
Stool 11/16 x 2 1/2"	5.65	2.65	3.90	6.20	4.22

		UNIT	Birch	Oak	Pine
.5	STAIRS (Treads, Risers, Skirt Boards)	LnFt	39.00	31.00	23.00
.6	SHELVING (12" Deep)	SqFt	42.00	37.00	25.00
.7	CUSTOM PANELING	SqFt	28.10	21.00	17.50
.8	THRESHOLDS - 3/4" x 3 1/2"	LnFt	-	20.00	-

0604.0   GLUE LAMINATE

	UNIT	COST
Arches 80' Span	SqFt	9.90
Beams & Purlins 40' Span	SqFt	6.80
Deck Fir - 3" x 6"	SqFt	8.50
4" x 6"	SqFt	9.40
Deck Cedar - 3" x 6"	SqFt	10.00
4" x 6"	SqFt	10.70
Deck Pine - 3" x 6"	SqFt	7.80

0605.0 PREFABRICATED WOOD & PLYWOOD COMPONENTS

			UNIT	COST		UNIT	COST
.1	WOOD TRUSSED RAFTERS - 24" O.C. - 4 -12 Pitch						
	Span To 16'	w/ Supports	Each	138.00	or	SqFt	4.30
	20'	and	Each	146.00		SqFt	3.65
	22'	2' Overhang	Each	149.00		SqFt	3.40
	24'		Each	155.00		SqFt	3.22
	26'		Each	161.00		SqFt	3.10
	28'		Each	178.00		SqFt	3.18
	30'		Each	210.00		SqFt	3.50
	32'		Each	233.00		SqFt	3.65
	Add for 5-12 Pitch		Each	5%		SqFt	15%
	Add for 6-12 Pitch		Each	12%		SqFt	25%
	Add for Scissor Truss		Each	30.00		SqFt	.23
	Add for Gable End 24'		Each	64.00		SqFt	.42
.2	WOOD FLOOR TRUSS JOISTS (TJI)						
	Open Web - Wood or Metal - 24" O.C.						
	Up To 23' Span x 12"	Single	LnFt	8.50	or	SqFt	4.24
	To 24' Span x 15"	Cord	LnFt	8.70		SqFt	4.35
	To 27' Span x 18"		LnFt	8.93		SqFt	4.46
	To 30' Span x 21"		LnFt	9.39		SqFt	4.70
	To 25' Span x 15"	Double	LnFt	8.35		SqFt	4.22
	To 30' Span x 18"	Cord	LnFt	8.70		SqFt	4.35
	To 36' Span x 21"		LnFt	9.70		SqFt	4.85
	Plywood Web - 24" O.C.						
	Up To 15' x 9 1/2"		LnFt	5.40		SqFt	2.70
	To 19' x 11 7/8"		LnFt	5.50		SqFt	2.73
	To 21' x 14"		LnFt	6.12		SqFt	3.05
	To 22' x 16"		LnFt	7.00		SqFt	3.40
.3	LAMINATED VENEER STRUCTURAL BEAMS						
.31	Micro Lam - Plywood - 24" O.C.						
	9 1/2" x 1 3/4"		LnFt	7.50	or	SqFt	3.80
	11 7/8" x 1 3/4"		LnFt	8.30		SqFt	4.10
	14" x 1 3/4"		LnFt	9.70		SqFt	4.85
	16" x 1 3/4"		LnFt	10.80		SqFt	5.40
.32	Glue Lam- Dimension Lumber- 24" O.C.						
	9" x 3 1/2"		LnFt	14.20		SqFt	7.10
	12" x 3 1/2"		LnFt	16.30		SqFt	8.20
	15" x 3 1/2"		LnFt	18.20		SqFt	9.10
	9" x 5 1/2"		LnFt	18.40		SqFt	9.20
	12" x 5 1/2"		LnFt	22.70		SqFt	11.30
	15" x 5 1/2"		LnFt	28.50		SqFt	14.25
	18" x 5 1/2"		LnFt	34.60		SqFt	17.30
	Add for Architectural Grade		LnFt	25%			
	Add for 6 3/4"		LnFt	40%			

0606.0 WOOD TREATMENTS

	PRESERVATIVES – PRESSURE TREATED			
	Dimensions		BdFt	.19
	Timbers		BdFt	.30
	FIRE RETARDENTS			
	Dimensions & Timbers		BdFt	.38

0607.0 ROUGH HARDWARE

.1	NAILS	SqFt or BdFt	.02
.2	JOIST HANGERS	Each	3.20
.3	BOLTS - 5/8" x 12"	Each	3.80

		UNIT	LABOR	MATERIAL
0701.0	**WATERPROOFING (L&M) (Roofers)**			
.1	MEMBRANE			
	1 Ply Membrane - Felt 15#	SqFt	.55	.40
	Fabric and Felt	SqFt	.76	.44
	2 Ply Membrane - Felt 15#	SqFt	.93	.63
	Fabric and Felt	SqFt	1.09	.66
.2	HYDROLITHIC	SqFt	.95	.94
.3	ELASTOMERIC- Rubberized Asphalt .060M & Poly	SqFt	.95	.64
	Liquid	SqFt	1.07	.77
.4	METALLIC OXIDE- 3 Coat	SqFt	.88	1.98
	3 Coat with Sand & Cement Cover	SqFt	1.13	2.20
.5	VINYL PLASTIC	SqFt	.55	1.27
.6	BENTONITE - 3/8"	SqFt	.77	.72
0702.0	**DAMPPROOFING (Roofers or Laborers)**			
.1	BITUMINOUS (Asphalt)			
.11	Trowel Mastics			
	Asphalt 1/16" (12 Sq.Ft. Gal.)	SqFt	.34	.25
	1/8" (20 Sq.Ft. Gal.)	SqFt	.39	.31
.12	Spray or Brush Liquid			
	Asphalt - 1 Coat - Spray	SqFt	.23	.17
	Spray	SqFt	.30	.22
	Brush	SqFt	.35	.24
.13	Hot Mop			
	Asphalt - 1 Coat - Including Primer	SqFt	.55	.42
	Fibrous Asphalt - 1 Coat	SqFt	.61	.45
.14	Trowel Mastics & Pre-formed Vapor Barriers			
	Mastic Fabric Mastic	SqFt	.57	.86
	and Polyvinyl Chloride	SqFt	.55	.72
.2	CEMENTITIOUS - 1 Coat - 1/2"	SqFt	.34	.18
	2 Coat - 1"	SqFt	.46	.22
.3	SILICONE (Masonry) - 1 Coat	SqFt	.25	.19
	2 Coat	SqFt	.33	.31
	Add for Scaffold to Above Sections	SqFt	.24	.13
0703.0	**BUILDING INSULATION (S) (Carpenters)**			
.1	FLEXIBLE (Batts and Rolls) - Friction			
	Fiberglass			
	2 1/4" R 7.4	SqFt	.27	.20
	3 1/2" R 11.0	SqFt	.28	.22
	3 5/8" R 13	SqFt.	.29	.30
	6" R 19.0	SqFt	.31	.35
	9" R 30.0	SqFt	.36	.60
	Add for Paper Faced Batts	SqFt	.02	.03
	Add for Polyethylene Barrier 2 Mil	SqFt	.04	.04
	Add for Staggered Stud Installation	SqFt	.10	.03

0703.0　BUILDING INSULATION, Cont'd...

		UNIT	LABOR	MATERIAL
.2	**RIGID**			
	Fiberglass			
	1" R 4.35 3# Density	SqFt.	.33	.36
	1 1/2" R 6.52 3# Density	SqFt	.35	.55
	2" R 8.70 3# Density	SqFt	.37	.72
	Add for 4.2# Density per Inch	SqFt	-	.08
	Add for 6.0# Density per Inch	SqFt	-	.30
	Add for Kraft Paper Backed	SqFt	-	.09
	Add for Aluminum Foil Backed	SqFt	-	.42
	Add for Fireproof Type	SqFt	-	10%
	Expanded Styrene			
	Molded, White (Bead Board)			
	1" x 16" x 8' R 5.40	SqFt	.33	.19
	1" x 24" x 8', T&G R 5.40	SqFt	.36	.26
	1 1/2" R 6.52	SqFt	.37	.29
	2" R 7.69	SqFt	.42	.37
	2" x 24" x 8', T&G R 7.69	SqFt	.42	.53
	Extruded, Blue			
	1" x 24" x 8' R 5.40	SqFt	.33	.43
	T&G	SqFt	.36	.54
	1 1/2" R 6.52	SqFt	.37	.54
	2" R 7.69	SqFt	.42	.56
	2" x 24" x 8', T&G R 7.69	SqFt	.42	.66
	Perlite			
	1" R 2.78	SqFt	.33	.41
	2" R 5.56	SqFt	.42	.83
	Urethane			
	1" R 6.67	SqFt	.35	.50
	2" R 13.34	SqFt	.42	.72
.3	**LOOSE**			
	Expanded Styrene 1" R 3.80	SqFt	.26	.24
	Fiberglass 1" R 2.20	SqFt	.28	.25
	Rockwool 1" R 2.90	SqFt	.29	.25
	Cellulose 1" R 3.70	SqFt	.26	.26
.4	**FOAMED**			
	Urethane per Inch or Bd.Ft.	SqFt	.50	.88
.5	**SPRAYED**			
	Fibered Cellulose per Inch or Bd.Ft.	SqFt	.46	.50
	Polystyrene per Inch or Bd.Ft.	SqFt	.55	.58
	Urethane per Inch or Bd.Ft.	SqFt	.60	1.05
.6	**ALUMINUM PAPER**	SqFt	.20	.18
.7	**VINYL FACED FIBERGLASS**			
	2 5/8" x 63 1/2" x 96" R 10.0	SqFt	.44	.94
	Add to All Above for Ceiling Work	SqFt	.07	-
	Add to All Above for Clip On	SqFt	.08	.02
	Add to All Above for Scaffold	SqFt	.22	.13

See Division 3 for Cementitious Poured and Rigid Insulation
See Division 4 for Core and Cavity Filled Masonry
See Division 7 for Roof Insulation
See Division 15 for Mechanical Insulations

		UNIT	COST
0704.0	**ROOF AND DECK INSULATION (L&M) (Roofers)**		
.1	RIGID - NON RATED		
	Fiber Board 1/2" R 1.39	Sq	56.20
	3/4" R 2.09	Sq	66.20
	1" R 2.78	Sq	76.95
	Fiberglass 1" R 3.70	Sq	88.70
	1 1/2" R 2.09	Sq	116.60
	Perlite 1/2" R 1.39	Sq	55.20
	3/4" R 2.09	Sq	68.40
	1" R 2.78	Sq	80.40
	2" R 5.26	Sq	120.00
	Expanded Polystrene 1" R 4.35	Sq	73.75
	1 1/2" R 6.52	Sq	86.00
	2" R 7.69	Sq	107.40
.2	RIGID - FIRE RATED		
	Isocyanurate 1" R 6.30 Glass Faced	Sq	122.00
	1 1/5" R 7.60	Sq	128.00
	1 1/2" R 10.00	Sq	134.30
	1 3/4" R 12.00	Sq	146.40
	2" R 14.00	Sq	151.60
	Phenolic Foam 1" R 7.60	Sq	101.30
	1 1/2" R 12.50	Sq	113.80
	2" R 16.60	Sq	140.00
	Add for 5/8" Sheet Rock over Steel Deck	Sq	16.20
	Add for Fiber or Mineral Cants	LnFt	1.80
	Add for Felt Faces	Sq	11.50
	Add for Composite Insulations	Sq	25%
.3	SPRAYED - 2" Polystyrene	Sq	275.00
0705.0	**MEMBRANE ROOFING (L&M) (Roofers)**		
.1	BUILT-UP ROOFING (50 SQUARES OR MORE)		
	Asphalt & Gravel - 15# Glass/Organic Felts 3-Ply	Sq	165.30
	4 - Ply	Sq	178.20
	5 - Ply	Sq	201.20
	Add for Pitch and Gravel	Sq	48.60
	Add for Curbed Openings	Each	144.30
	Add for Round Vent Openings	Each	130.70
	Add for Roof Skids	LnFt	31.50
	Add for Sloped Roofs	Sq	37.10
	Add for Applications on Wood	Sq	12.65
	Add for Marble Chips	Sq	55.00
	Add for Barriers - 15" Felt	Sq	10.90
	Fire Resistant	Sq	24.90
	Add for Walkways - 1"	SqFt	5.30
	Add for Bond	Sq	12.00
	Add for Roof Openings Cut in Existing	Each	459.00

See Quick Estimating 7A-10 for Combined Roofing, Insulation and Sheet Metal

0706.0 SINGLE-PLY ROOFING No Insulation or Flashing Included

	UNIT	Gravel Ballast	Application Mechanical Fastened	Adhesive Fastened
Butylene 100 mil	Sq	204.00	220.00	225.00
EPDM 50 mil	Sq	160.00	193.00	220.00
Neoprene 60 mil	Sq	297.00	358.00	248.00
P.V.C. 50 mil	Sq	149.00	215.00	237.00
Add per Opening	Opng	143.00	-	-

0707.0 TRAFFIC ROOF COATINGS (L&M)

	UNIT	LABOR	MATERIAL
.1 PEDESTRIAN TRAFFIC TYPE	SqFt	-	7.15
.2 VEHICULAR TRAFFIC TYPE			
Rubberized Coating with 3/4" Asphalt Topping	SqFt	-	4.80
Polyurethane with Non-Slip Aggregates on 20 mil silicone rubber	SqFt	-	4.40
Elastomeric with Non-Skid - 2 Coat	SqFt	-	2.90
3 Coat	SqFt	-	3.70

0708.0 SHINGLE, TILE and CORRUGATED ROOFING

	UNIT	LABOR	MATERIAL
.1 ASPHALT SHINGLES (S) (Carpenters) 4-12 Pitch			
Square Butt 235#- Seal Down	SqFt	.47	.44
300# - Laminated	SqFt	.52	.66
325# - Fire Resistant	SqFt	.55	.68
Timberline 300# - 25 Year	SqFt	.63	.57
300# - 30 Year	SqFt	.65	.72
Tab Lock 340#	SqFt	.63	.99
Roll 90# - 100 SqFt per Roll	SqFt	.29	.21
55# - 100 SqFt per Roll	SqFt	.24	.18
Base Starter - 240 SqFt per Roll	SqFt	.24	.25
Ice & Water Starter - 100 SqFt per Roll	SqFt	.30	.46
Add for Boston Ridge Shingle	LnFt	1.20	.53
Add for 15# Felt Underlayment	SqFt	.06	.04
Add for Pitches 6-12 & Up - Ea Pitch Increase	SqFt	.06	-
Add for Removal and Haul - Away (2 Layers)	SqFt	.32	.15
Add for Chimneys, Skylights & Bay Windows	Each	47.00	-
.2 FIBERGLASS SHINGLES - 215# - Seal Down	SqFt	.47	.44
250# - Seal Down	SqFt	.50	.63
.3 WOOD			
Sawed			
Red Cedar - 16" x 5" to Weather - #1 Grade	SqFt	.84	1.70
#2 Grade	SqFt	.82	1.30
#3 Grade	SqFt	.76	.80
Hand Splits			
Red Cedar - 1/2" x 3/4", 24" x 10" to Weather	SqFt	.99	1.40
3/4"x1 1/4", 24" x 10" to Weather	SqFt	1.07	1.70
Fancy Butt - 7" to Weather	SqFt	1.20	3.60
Add for Fire Retardant	SqFt	-	.44
Add for Pitches over 5-12, Each Pitch Increase	SqFt	.08	-
.4 METAL			
Aluminum - Mill .020	SqFt	.88	1.70
.030	SqFt	.89	2.05
Anodized .020	SqFt	.88	2.13
.030	SqFt	.89	2.42
Steel - Enameled - Colored	SqFt	.93	2.95
Galvanized - Colored	SqFt	.93	2.24

0708.0 SHINGLE, TILE AND CORRUGATED ROOFING, Cont'd...

			UNIT	COST
.6	SLATE SHINGLES (Incl. Underlayment) (L&M) (Roofers)			
	Vermont and Pennsylvania			
	3/16" - 6 1/2" to 8 1/2" to Weather			
		Weathered Green or Black	Sq	1,113.00
		Gray	Sq	1,113.00
		Fading Purple	Sq	1,113.00
		Unfading Purple	Sq	1,452.00
		Add for Variegated Colors & Sizes - 1/4" to 1/2"	Sq	376.00
		3/8" to 1"	Sq	720.00
		Add for Hips and Valleys	LnFt	56.00
		Add for Pitches over 6-12	Sq	76.00
.7	TILE (L&M) (Roofers)			
.71	Clay, Interlocking Shingle Type - Designed		Sq	1,097.00
	Early American		Sq	1,102.00
	Williamsburg		Sq	1,219.00
	Mormon		Sq	1,760.00
	Architectural Pattern Shingle Type - Provincial		Sq	1,050.00
	Georgian		Sq	1,050.00
	Colonial		Sq	1,050.00
	Spanish, Red		Sq	753.00
	Mission		Sq	1,219.00
.72	Concrete - Plain - 10" to Weather - 9" Wide		Sq	668.00
	Colored		Sq	742.00
	Add for Staggered Installation		Sq	244.00
.73	Steel - 16 Ga - Ceramic Coated		Sq	455.00
.74	Aluminum .019		Sq	530.00
.8	CORRUGATED PANELS			
.81	Concrete - Plain		Sq	3.80
.82	Vinyl - 120		Sq	6.00
.83	Fiberglass - 8 oz		Sq	4.00
.84	Aluminum .024 - Plain		Sq	3.35
	.024 - Painted		Sq	4.10
	See Division 5 for Corrugated Metal Panels			

0709.0 SHEET METAL WORK (L&M)

(Sheet Metal Workers)	UNIT	Galvanized 25 Ga	Aluminum .032"	Copper 16 oz	Enameled
.1 DOWNSPOUTS, GUTTERS AND FITTINGS					
Downspouts - 3" x 2"	LnFt	2.65	2.65	7.25	3.31
4" x 3"	LnFt	3.30	3.56	10.30	4.43
5" x 4"	LnFt	4.54	4.60	11.30	5.67
Round 3"	LnFt	2.54	3.13	6.68	3.30
4"	LnFt	3.00	3.73	7.90	4.19
5"	LnFt	3.78	4.53	8.82	5.02
Chute Type 4" x 6"	LnFt	4.38	5.10	-	5.05
5" x 7"	LnFt	5.01	5.72	-	6.96
Gutters - Box or Half Round 4"	LnFt	3.50	4.54	7.24	5.31
5"	LnFt	4.09	4.64	8.94	5.28
6"	LnFt	5.01	5.33	9.15	5.59
Add for Guards	LnFt	-	1.12	-	-
Fittings (Elbows, Corners, etc.)	LnFt	10.55	9.96	12.40	12.40

See Quick Estimating 7A-10 for Combined Roofing, Insulation and Sheet Metal.

0709.0 SHEET METAL WORK, Cont'd...

.2 FASCIAS, GRAVEL STOPS & TRIM

	UNIT	COST				
		Galva-nized 25 Ga	Alum-inum .032"	Copper 16 oz	Enameled	PVC
Fascias - 6"	LnFt	5.00	7.70	11.00	6.25	6.95
8"	LnFt	5.34	8.61	12.30	6.58	7.60
Copings - 12"	LnFt	6.30	10.14	12.80	7.70	-
Scuppers	Each	119.00	119.00	142.00	124.00	-
Gravel Stops - 4"	LnFt	4.90	7.15	8.40	6.20	-
6"	LnFt	4.54	8.03	10.40	6.39	7.17
8"	LnFt	5.90	8.50	11.80	7.20	7.65
Add for Durodonic Finish	LnFt		25%			

.3 FLASHINGS

	UNIT	COST
Galvanized - 25 ga	SqFt	5.00
Aluminum - .019	SqFt	4.70
.032	SqFt	5.00
Copper - 16 oz	SqFt	8.10
20 oz	SqFt	8.30
Asphalted Fabric - Plain	SqFt	2.50
Copper Backed	SqFt	4.07
Aluminum Back	SqFt	3.58
Vinyl Fabric - .020	SqFt	1.30
.030	SqFt	1.50
Lead - 2.5#	SqFt	7.55
Rubber - 1/16"	SqFt	8.30
PVC - 24 mil	SqFt	3.40

.4 SHEET METAL ROOFING - Including Underlayment

	UNIT	COST
Copper - Standing Seam - 16 oz	Sq	1,160.00
20 oz	Sq	1,190.00
Batten Seam 16 oz	Sq	1,055.00
20 oz	Sq	1,225.00
Flat Lock 16 oz	Sq	975.00
20 oz	Sq	1,125.00
Stainless Steel - Batten Seam 28 ga	Sq	1,070.00
Standing Seam 28 ga	Sq	1,190.00
Flat Seam 28 ga	Sq	1,010.00
Monel - Batten Seam .018"	Sq	1,203.00
Standing Seam .018"	Sq	1,240.00
Flat Seam .108"	Sq	1,145.00
Lead - Copper Coated - Batten Seam 3 lb	Sq	1,135.00
Flat Seam 3 lb	Sq	1,065.00
Aluminum - Clad	Sq	875.00
Galvanized - Colored .020"	Sq	954.00

See Division 5 for Corrugated Type

.5 EXPANSION JOINTS - 2 1/2"

	UNIT	COST
Polyethylene with galvanized	LnFt	7.30
Galvanized 25 ga	LnFt	8.51
Aluminum .032	LnFt	9.65
Copper 16 oz	LnFt	23.40
Stainless Steel	LnFt	12.80
Neoprene	LnFt	9.34
Butyl	LnFt	7.50
Add for 3 1/2"	LnFt	10%

0710.0	ROOF AND SOFFITT ACCESSORIES			or		
	(L&M) (Roofers)	UNIT	COST	or	UNIT	COST
.1	SKYLIGHTS					
	Example: 4' - 0" x 4' - 0"	SqFt	37.80		Each	636.00
	8' - 0" x 8' - 0"	SqFt	23.70		Each	1,548.00
.2	SKY DOMES (Plastic - Clear or White)					
	20" x 20" Single				Each	175.00
	Double				Each	207.00
	24" x 24" Single				Each	212.00
	Double				Each	244.00
	48" x 48" Single				Each	514.00
	Double				Each	583.00
	72" x 72" Single				Each	864.00
	Double				Each	1,039.00
	96" x 96" Double				Each	1,643.00
	Add for Bronze Top Colors				-	10%
	Add for Electric Operated Type				Each	4.60
	Add for Pyramid Type				Each	100%
	Add for Hip Type				Each	50%
.3	GRAVITY VENTILATORS					
	Diameter Aluminum or Galvanized 12"				Each	111.00
	24"				Each	265.00
.4	SMOKE VENTILATORS					
	Example: 48" x 96" - Aluminum				Each	2,268.00
	48" x 48" - Steel				Each	2,120.00
.5	PREFAB CURBS & EXPANSION JOINTS - 3" Fibered				LnFt	1.75
	4" Fibered				LnFt	1.80
.6	ROOF LINE LOUVRES				LnFt	1.70
.7	TURBINE VENTILATORS - 12"				Each	345.00
	24"				Each	355.00
.8	CUPOLAS				Each	366.00
.9	ROOF VENTS - P.V.C.				Each	64.00
	Aluminum				Each	58.00
	See 1012 for Roof Hatches					
0711.0	**SEALANTS (L&M) (Carpenters)**					
.1	CAULKING - 1/2" x 1/2" Joints					
	Windows and Doors - 2-Part Proprietary (Oil Base)				LnFt	2.00
	Silicone Rubber				LnFt	2.33
	Polysulfide (Thiakol)				LnFt	2.51
	Acrylic				LnFt	2.44
	Polyurethane				LnFt	2.44
	Control Joints - 2 - Part Proprietary				LnFt	2.20
	Silicone Rubber				LnFt	2.61
	Polysulfide				LnFt	3.56
	Acrylic				LnFt	2.65
	Polyurethane				LnFt	2.65
	Stone Pointing -Silicone Rubber				LnFt	3.00
	Polysulfide				LnFt	3.20
	Acrylic				LnFt	2.61
	Add for Swing Stage Work				LnFt	.50
	Add for 3/4" x 3/4" Joints				LnFt	.1.00
	Deduct for 1/4" x 1/4" Joints				LnFt	.30
.2	GASKETS AND JOINT FILLERS					
	Gaskets - 1/4" Polyvinyl x 6"				LnFt	2.40
	1/2" Polyvinyl x 6"				LnFt	3.00
	1/4" Neoprene x 6"				LnFt	3.20
	1/2" Neoprene x 6"				LnFt	4.10
	Joint Fillers -1/2"				LnFt	.37
	3/4"				LnFt	.42
	1"				LnFt	.45

	SQ.FT. COST
0701.0 WATERPROOFING	
1-Ply Membrane Felt 15#	1.49
2-Ply Membrane Felt 15#	2.30
Hydrolithic	2.50
Elastomeric Rubberized Asphalt with Poly Sheet	2.30
Liquid	2.50
Metallic Oxide 3-Coat	3.80
Vinyl Plastic	2.60
Bentonite 3/8" - Trowel	1.80
5/8" - Panels	2.05
0702.0 DAMPPROOFIN	
Asphalt Trowel Mastic 1/16"	.91
1/8"	1.08
Spray Liquid 1-Coat	.57
2-Coat	.79
Brush Liquid 2-Coat	.90
Hot Mop 1-Coat and Primer	1.50
1 Fibrous Asphalt	1.60
Cementitious Per Coat 1/2" Coat	.73
Silicone 1-Coat	.54
2-Coat	.83
Add for Scaffold and Lift Operations	.49
0703.0 BUILDING INSULATION	
FLEXIBLE	
Fiberglass 2 1/4" R 7.40	.71
3 1/2" R 11.00	.76
3 5/8" R 13.00	.89
6" R 19.00	.99
9" R 30.00	1.30
12" R 38.00	1.40
Add for Ceiling Work	.09
Add for Paper Faced	.10
Add for Polystyrene Barrier (2m)	.09
Add for Scaffold Work	.54
RIGID	
Fiberglass 1" R 4.35 3# Density	.99
1 1/2" R 6.52 3# Density	1.24
2" R 8.70 3# Density	1.50
Styrene, Molded 1" R 4.35	.78
1 1/2" R 6.52	.91
2" R 7.69	1.17
2" T&G R 7.69	1.49
Styrene, Extruded 1" R 5.40	1.09
1 1/2" R 6.52	1.27
2" R 7.69	1.38
2" T&G R 7.69	1.51
Perlite 1" R 2.78	1.08
2" R 5.56	1.71
Urethane 1" R 6.67	1.21
2" R 13.34	1.59
Add for Glued Applications	.09

	SQ.FT.COST
0703.0 BUILDING INSULATION, Cont'd...	
LOOSE 1" Styrene R 3.8	.76
1" Fiberglass R 2.2	.73
1" Rock Wool R 2.9	.75
1" Cellulose R 3.7	.72
FOAMED Urethane Per Inch	1.87
SPRAYED Cellulose Per Inch	1.36
Polystyrene	1..60
Urethane	2.30
ALUMINUM PAPER	.58
VINYL FACED FIBERGLASS	1.80

0704.0 ROOF & DECK INSULATION - COMBINED WITH 0709.0 BELOW

0705.0 MEMBRANE ROOFING - COMBINED WITH 0709.0 BELOW

0708.0 SHINGLE ROOFING

ASPHALT SHINGLES 235# Seal Down	1.34
300# Laminated	1.65
325" Fire Resistant	1.71
Timberline	1.68
340# Tab Lock	2.20
ASPHALT ROLL 90#	.78
55#	.70
FIBERGLASS SHINGLES 215#	1.34
250#	1.60
RED CEDAR 16" x 5" to Weather #1 Grade	3.60
#2 Grade	3.10
#3 Grade	2.31
24" x 10" to Weather - Hand Splits - 1/2" x 3/4" #2	3.36
3/4" x 1 1/4" #2	3.90
Add for Fire Retardant	.56
METAL Aluminum 020 Mill	3.60
020 Anodized	4.20
Steel, Enameled Colored	5.30
Galvanized Colored	4.30
Add to Above for 15# Felt Underlayment	
Add for Base Starter	.69
Add for Ice & Water Starter	1.08
Add for Boston Ridge	2.40
Add for Removal & Haul Away	.90
Add for Pitches Over 5-12 each Pitch Increase	.06
Add for Chimneys, Skylights and Bay Windows - Each	70.00

0704.0/ 0705.0/ 0709.0 - COMBINED ROOFING INSULATION and SHEET METAL

MEMBRANE	SQ. COST
3-Ply Asphalt & Gravel - Fiberboard & Perlite - R 10.0	464.00
R 16.6	518.00
4-Ply - R 10.0	529.00
R 16.6	556.00
5-Ply - R 10.0	610.00
R 16.6	675.00
Add for Sheet Rock over Steel Deck - 5/8"	124.00
Add for Fiberglass Insulation	41.00
Add for Pitch and Gravel	63.00
Add for Sloped Roofs	51.00
Add for Thermal Barrier	40.00
Add for Upside Down Roofing System	116.00
SINGLE-PLY - 60M Butylene Roofing & Gravel Ballast - R 10.0	437.00
Mech. Fastened - R 10.0	448.00
PVC & EPDM Roofing & Gravel Ballast - R 10.0	416.00
Mech. Fastened - R 10.0	448.00
Blocking and Cants Not Included	

0801.0 HOLLOW METAL (M) (CARPENTERS)

	UNIT	LABOR	MATERIAL
.1 CUSTOM			
.11 Frames (16 Gauge)			
2'6" x 6'8" x 4 3/4"	Each	54	170
2'8" x 6'8" x 4 3/4"	Each	55	180
3'0" x 6'8" x 4 3/4"	Each	56	191
3'4" x 6'8" x 4 3/4"	Each	66	196
3'6" x 6'8" x 4 3/4"	Each	74	201
4'0" x 6'8" x 4 3/4"	Each	85	212
Deduct for 18 Gauge	Each	8	18.55
Add for 14 Gauge	Each	13	28
Add for A, B, or C Label	Each	8	14.05
Add for Transoms	Each	34	85
Add for Side Lights	Each	36	104
Add for Frames over 7'0" (Height)	Each	36	18.55
Add for Frames over 6 3/4" (Width)	Each	14	18.55
Add for Bolted Frame with Bolts	Each	25	38
.12 Doors (1 3/4" - 18 Gauge)			
2'6" x 6'8" x 1 3/4"	Each	57	207
2'8" x 6'8" x 1 3/4"	Each	58	217
3'0" x 6'8" x 1 3/4"	Each	60	228
3'4" x 6'8" x 1 3/4"	Each	69	239
3'6" x 6'8" x 1 3/4"	Each	78	260
4'0" x 6'8" x 1 3/4"	Each	90	292
Add for 16 Gauge	Each	12	43
Deduct for 20 Gauge	Each	8	18.55
Add for A Label	Each	10	18.55
Add for B or C Label	Each	8	18.55
Add for Vision Panels or Lights	Each	-	82
Add for Louvre Openings	Each	-	76
Add for Galvanizing	Each	-	50
.2 STOCK			
.21 Frames (4 3/4"-5 3/4"-6 3/4"-8 3/4"- 16 Ga)			
2'6" x 6'8" or 7'0"	Each	53	101
2'8" x 6'8" or 7'0"	Each	54	109
3'0" x 6'8" or 7'0"	Each	57	114
3'4" x 6'8" or 7'0"	Each	64	120
3'6" x 6'8" or 7'0"	Each	72	127
4'0" x 6'8" or 7'0"	Each	82	154
See above .11 for Size Change			
Add for Bolted Frame & Bolts	Each	22	38
Add for A, B, or C Label	Each	8	18.55
Deduct Non-Welded Frame (Knock Down)	Each	6	18.55
.22 Doors (1 3/8" or 1 3/4" - 18 Gauge)			
2'6" x 6'8" or 7'0"	Each	57	212
2'8" x 6'8" or 7'0"	Each	58	217
3'0" x 6'8" or 7'0"	Each	60	228
3'4" x 6'8" or 7'0"	Each	68	239
3'6" x 6'8" or 7'0"	Each	76	276
4'0" x 6'8" or 7'0"	Each	84	286
Add for A Label	Each	8	18.55
Add for B or C Label	Each	7	13.00
Add for Galvanizing	Each	-	50

Labor Above Includes Installation of Butts & Locksets

Add to All Above for:

	UNIT	LABOR	MATERIAL
Door Closures - Surface Mounted	Each	52	113
Panic Bars (Exit Device)	Each	11	525
Kick, Push and Pull Plates	Each	22	37

See 8-13 for Other Hardware

0802.0   WOOD DOORS (M) (CARPENTERS)
See 0603.3 and 0802.5 for Custom Frames
Labor includes Installation of Butts and Locksets

.1   FLUSH DOORS - Pre-Machined
No Label - Standard - Birch - 7 Ply - Paint Grade

	UNIT	LABOR	Hollow Core	Solid Core
2'0" x 6'8" or 1 3/8"	Each	55	110	121
2'4" x 6'8" or 1 3/8"	Each	56	116	126
2'6" x 6'8" or 1 3/8"	Each	57	121	131
2'8" x 6'8" or 1 3/8"	Each	58	126	137
3'0" x 6'8" or 1 3/8"	Each	60	131	147
2'0" x 6'8" or 1 3/4"	Each	56	116	131
2'4" x 6'8" or 1 3/4"	Each	57	121	142
2'6" x 6'8" or 1 3/4"	Each	58	126	147
2'8" x 6'8" or 1 3/4"	Each	59	126	152
3'0" x 6'8" or 1 3/4"	Each	64	131	158
3'4" x 6'8" or 1 3/4"	Each	68	152	179
3'6" x 6'8" or 1 3/4"	Each	72	231	262
3'8" x 6'8" or 1 3/4"	Each	78	252	273
Add for Stain Grade	Each			42
Add for 5 Ply	Each			84
Add for Jamb & Trim - Solid - Knock Down	Each			84
Add for Jamb & Trim - Veneer - Knock Down	Each			74
Add for Red Oak - Rotary	Each			24
Add for Red Oak - Plain Sliced	Each			42
Add for Architectural Grade - 7 Ply	Each			89
Deduct for Lauan	Each			16
Add for Doors over 6'8" per Inch	Each			16
Add for Vinyl Overlay	Each			49
Add for Cutouts and Special Cuts	Each			57
Add for Lite Cutouts with Metal Frames	Each			83
Add for Private Door Eye	Each			23
Add for Wood Louvre	Each			116
Add for Transom Panels & Side Panels	Each	28		126
Add for Astragals	Each	29		55
Add for Trimming Door for Carpet	Each	24		-

Label - Birch - Paint Grade

	UNIT	LABOR	20-Min	45-Min	60-Min
				MATERIAL	
2'0" x 6'8" x 1 3/4"	Each	55	147	179	189
2'4" x 6'8" x 1 3/4"	Each	56	158	189	205
2'6" x 6'8" x 1 3/4"	Each	57	163	200	210
2'8" x 6'8" x 1 3/4"	Each	58	168	247	263
3'0" x 6'8" x 1 3/4"	Each	59	173	289	294
3'4" x 6'8" x 1 3/4"	Each	67	189	310	305
3'6" x 6'8" x 1 3/4"	Each	72	273	326	326
3'8" x 6'8" x 1 3/4"	Each	80	299	336	347

See Above 0602.1 for Changes

.2   PANEL DOORS

	UNIT	LABOR	Pine	Fir	Oak
				MATERIAL	
Exterior Entranc					
2'8" x 6'8" x 1 3/4"	Each	85	368	378	588
3'0" x 6'8" x 1 3/4"	Each	90	389	399	609
Add for Sidelights	Each	42	215	236	378
Inter					
2'6" x 6'8" x 1 3/8"	Each	64	326	305	483
2'8" x 6'8" x 1 3/8"	Each	69	336	315	494
3'0" x 6'8" x 1 3/8"	Each	74	368	326	504

0802.0 WOOD DOORS (Cont'd...)

		UNIT	LABOR	Birch/Oak	Pine	Lauan
				MATERIAL		
.3	LOUVERED DOORS					
	1'0" x 6'8" x 1 3/8" Pine	Each	37	420	158	-
	1'3" x 6'8" x 1 3/8"	Each	40	452	163	-
	1'6" x 6'8" x 1 3/8"	Each	45	462	168	-
	2'0" x 6'8" x 1 3/8"	Each	54	546	194	-
	2'4" x 6'8" x 1 3/8"	Each	59	546	210	-
	2'6" x 6'8" x 1 3/8"	Each	59	557	263	-
	2'8" x 6'8" x 1 3/8"	Each	61	578	273	-
	3'0" x 6'8" x 1 3/8"	Each	74	59	-	
	Add for Half Louvered/Panel	Each				15%
	Add for Prehung	Each	-	100	100	0
.4	BI-FOLD - FLUSH					
	2'0" x 6'8" x 1 3/8" Pine	Each	54	-	-	92
	2'4" x 6'8" x 1 3/8"	Each	57	-	-	100
	2'6" x 6'8" x 1 3/8"	Each	61	-	-	100
	2'8" x 6'8" x 1 3/8"	Each	64	-	-	105
	3'0" x 6'8" x 1 3/8"	Each	69	-	-	113
	Add for Oak	Each	-	41	-	-
	Add for Birch	Each	-	36	-	-
	Add for Louvered Type	Each	11			89
	Add for Half Louvered/Panel	Each	11			89
	Add for Decorative Type	Each	11			184
	Add for Prefinished Type	Each	-			23
	Add for Heights over 6'8"	Each	11			5
	Add for Widths over 3'0"	Each	11			6
	See 0803.2 for Closet Bi-fold Doors and Prehung Stock					

		UNIT	LABOR	Oak	Pine	Fir
.5	CAFE DOORS					
	2'6" x 3'8" x 1 1/8"	Pair	72	-	263	273
	2'8" x 3'8" x 1 1/8"	Pair	80	-	284	284
	3'0" x 3'8" x 1 1/8"	Pair	85	-	294	294
.6	FRENCH - with Grille Patte					
	2'6" x 6'8" x 1 3/8"	Each	85	525	599	368
	2'8" x 6'8" x 1 3/8"	Each	87	536	609	399
	3'0" x 6'8" x 1 3/8"	Each	90	546	620	546
	Add for Prehung	Each	-	100	-	-
.7	DUTCH					
	2'6" x 6'8" x 1 3/4"	Each	108	-	735	-
	2'8" x 6'8" x 1 3/4"	Each	113	-	814	-
	3'0" x 6'8" x 1 3/4"	Each	119	-	840	-

.8 PREHUNG DOOR UNITS - Incl. Frame, Trim, Hardware and Jambs

	UNIT	LABOR	Birch	Pine	Fir
Exterior					
2'8" x 6'8" x 1 3/4"	Each	72	-	336	-
3'0" x 6'8" x 1 3/4"	Each	80	-	389	-
3'4" x 6'8" x 1 3/4"	Each	85	-	494	-
Add for Insulated	Each	11	-	-	147
Add for Threshold	Each	11	-	-	19
Add for Weatherstrip	Each	-	-	-	91
Add - Sidelites 1'0" HC	Each	25	-	-	173
Add - Sidelites 1'6" HC	Each	30	-	-	210

	UNIT	LABOR	Birch	Lauan	Oak
Interior					
2'6" x 6'8" x 1 3/8"	Each	67	226	158	231
2'8" x 6'8" x 1 3/8"	Each	71	236	163	236
3'0" x 6'8" x 1 3/8"	Each	78	247	168	247

0803.0 SPECIAL DOORS

					MATERIAL		
		UNIT	LABOR	Oak Flush	Birch Flush	Lauan Flush	Pine & Fir Panel
.1	BLAST OR NUCLEAR RESISTANT						
.2	CLOSET - BI-FOLDING (PREHUNG STOCK)						
.21	Wood - Units - 1 3/8"						
	Incl. Jamb, Csg. Track & Hdwe.						
	2 Door - 2'0" x 6'8"	Each	49	184	200	131	441
	2'6" x 6'8"	Each	53	191	210	137	472
	3'0" x 6'8"	Each	57	200	221	144	525
	4 Door - 4'0" x 6'8"	Each	64	300	294	205	735
	5'0" x 6'8"	Each	71	310	315	210	840
	6'0" x 6'8"	Each	80	331	341	236	908
	Incl. Track & Hdwe (Unfinished)						
	2 Door - 2'0" x 6'8"	Each	41	110	103	103	-
	2'6" x 6'8"	Each	47	121	109	110	-
	3'0" x 6'8"	Each	49	131	121	121	-
	4 Door - 4'0" x 6'8"	Each	52	179	158	158	-
	5'0" x 6'8"	Each	55	192	188	173	-
	6'0" x 6'8"	Each	59	215	210	189	-
	Add for Louvered Type	Each					100%

		UNIT	LABOR	MATERIAL
.22	Metal			
	Flush 2 Door - 2'0" x 6'8"	Each	40	95
	2'6" x 6'8"	Each	42	100
	3'0" x 6'8"	Each	45	105
	Flush 4 Door - 4'0" x 6'8"	Each	49	131
	5'0" x 6'8"	Each	51	137
	6'0" x 6'8"	Each	57	147
	Add for Plastic Overlay	Each	-	34
	Add for Louvered or Decorative	Each	-	11
.23	Leaded Mirror			
	4 Panel Unit - 4'0" x 6'8"	Each	89	625
	5'0" x 6'8"	Each	95	683
	6'0" x 6'8"	Each	106	735
.3	FLEXIBLE DOORS (M) (CARPENTER)			
	(See 1020 - Retractable Partitions)			
	Wood Slat (Unfinished)	SqFt	2.12	12.10
	Fabric Accordion Fold	SqFt	2.12	10.00
	Vinyl Clad	SqFt	2.12	13.65

				COST
.4	GLASS-ALL (L&M) (GLAZIERS)			
	Manual (Including Hardware)	Each	-	2,310
	Add for Center Locking	Each	-	184
	Automatic (Including Hardware & Operator)			
	Mat or Floor Unit	Each	-	6,615
	Handle Unit	Each	-	7,140
.5	HANGAR (L&M) (IRONWORKERS)			
	Biparting or Sliding - Steel			
	Example: 100' x 20'	Each	-	57,750
	Add for Pass Doors	Each	-	1,418
	Add for Electric Operator	Each	-	1,838
	Add for Insulating	Each	-	2,048
	Tilt Up or Canopy	Each		Approx. Same
.6	METAL COVERED FIRE & INDUSTRIAL SLIDING DOORS			
	3'0" x 7'0" - Flush	Each	-	714
	4'0" x 7'0"	Each	-	2,048
	8'0" x 8'0" - U.L. Label	Each	-	2,730
	10'0" x 10'0"	Each	-	3,465
	10'0" x 12'0"	Each	-	3,780
	12'0" x 12'0"	Each	-	4,200

0803.0   SPECIAL DOORS, Cont'd...

.7   OVERHEAD DOORS (L&M) (CARPENTERS)

	UNIT	1 3/4" Wood	2" 24-Ga Steel
Commercial			
8' x 8'	Each	893.00	903.00
8' x 10'	Each	1,050.00	1,074.00
8' x 12'	Each	1,155.00	1,208.00
10' x 10'	Each	1,313.00	1,418.00
10' x 12'	Each	1,418.00	1,470.00
12' x 12'	Each	1,575.00	1,628.00
14' x 10'	Each	1,628.00	1,680.00
14' x 12'	Each	1,995.00	2,048.00
Add for Panel Door	Each		5%
Add for Low Lift	Each		82.00
Add for High Lift	Each		137.00
Add for Electric Operator	Each		604.00
Add for Insulated Type	Each		10%
Residential			
8' x 7' x 1 3/4"	Each	494.00	504.00
9' x 7' x 1 3/4"	Each	525.00	525.00
16' x 7' x 1 3/4"	Each	1,071.00	1,155.00
Add for Wood Paneled Door	Each		10%
Add for Redwood	Each		236.00
Add for Electric Operator	Each		446.00
Add for Vision Lights	Each		23.00

.8   PLASTIC LAMINATE FACED (M) (CARPENTERS)

	UNIT	LABOR	MATERIAL
1 3/8"	SqFt	4.55	13.00
1 3/4"	SqFt	4.65	13.00
Add for Edge Strip - Vertical	Each		7.60
Add for Edge Strip - Horizontal	Each		6.60
Add for Cutouts	Each		63.00
Add for Metal Frame Cutout	Each		91.00
Add for Decorative Type	SqFt		20.00
Add for Solid Colors	SqFt		10%
Add for 1-hour B Label	SqFt		5.40
Add for 1 1/2-hour B Label	SqFt		5.50
Add for Color Core	-		50%
Add for Machining for Hardware	Each		53.00

.9   REVOLVING DOOR (L&M) (GLAZIERS)

	UNIT	COST
Aluminum		
3 Leaf - 6'-6" x 6'-6"	Each	23,625.00
4 Leaf - 6'-6" x 6'-6"	Each	25,200.00
6'-0" x 7'-0"	Each	18,375.00
Automatic - 6'-6" x 10'-0"	Each	35,700.00
Stainless	Each	30,975.00
Bronze	Each	34,650.00
Painted Steel - Dark Room 32"	Each	2,048.00
36"	Each	2,100.00

0803.0 SPECIAL DOORS, Cont'd...	UNIT	LABOR	MATERIAL
.10 ROLLING DOORS & GRILLES (M) (IRONWORKERS)			
Doors - Manual - 8' x 8'	Each	435.00	2,067.00
10' x 10'	Each	414.00	2,120.00
12' x 12'	Each	404.00	2,862.00
14' x 14'	Each	784.00	4,028.00
Add for Electric Controlled	Each	-	318.00
Add for Fire Doors	Each	-	700.00
Grilles - Manual -6'8" x 3'2"	Each	191.00	848.00
6'8" x 4'2"	Each	175.00	943.00
8'0" x 3'2"	Each	212.00	996.00
8'0" x 4'2"	Each	223.00	1,113.00
Add for Electric Controlled	Each	-	297.00
.11 SHOWER DOORS (M) (CARPENTERS)			
28" x 66" Obscure Pattern Glass Wire	Each	60.00	159.00
Tempered Glass	Each	60.00	170.00
.12 SLIDING OR PATIO DOORS (M) (CARPENTERS)			
ONE SLIDING, ONE FIXED			
Metal (Aluminum) (Including 5/8" Glass, Threshold, Hardware and Screen)			
8'0" x 6'10"	Each	159.00	1,246.00
8'0" x 8'0"	Each	212.00	1,378.00
Wood (M) (Carpenters) Vinyl Sheathed (Including 5/8" Insulated Dbl. Temp. Glass, Weatherstripped, Hardware and Casing)			
Vinyl Clad 6' 0" x 6' 8"	Each	159.00	1,696.00
8' 0" x 6' 8"	Each	191.00	2,014.00
Pine - Prefinished 6' 0" x 6' 8"	Each	159.00	1,696.00
8' 0" x 6' 8"	Each	189.00	1,908.00
Deduct for 3/16" Safety Glass/Insulated	Each	-	186.00
Add for Tripe Glazing	Each	-	223.00
Add for Screen	Each	17.00	154.00
Add for Grilles	Each	21.00	233.00
.13 SOUND REDUCTION (M) (CARPENTERS)			
Metal	Each	175.00	1,802.00
Wood	Each	163.00	954.00
.14 TRAFFIC (M) (CARPENTERS)			
Electric Powered			
Hollow Metal - 8' x 8'	Each	440.00	4,982.00
8' x 8' High Speed	Each	1,113.00	11,130.00
Metal Clad (26 ga.) - 8' x 8'	Each	413.00	4,770.00
Wood	Each	360.00	3,922.00
Truck Impact - Double Acting (Including Bumpers and Vision Panels)			
Metal Clad - 8' x 8' Openings	Each	313.00	2,756.00
Rubber Plastic - 8' x 8' Openings 1/2"	Each	249.00	2,279.00
Aluminum - 8' x 8' Openings 1 3/4"	Each	270.00	2,862.00
.15 SPECIAL MADE & ENGINEERED INDUSTRIAL DOORS (L&M) (CARPENTERS)			
Four-Fold Garage Doors - Electric Operated	SqFt	-	111.00
.16 VAULT DOORS (M) (STEEL ERECTORS)			
1 hour with Frame 6' 6" x 2' 8"	Each	366.00	2,332.00
2 hour with Frame 6' 6" x 2' 8"	Each	366.00	3,074.00
6' 6" x 3' 0"	Each	440.00	3,498.00
6' 6" x 3' 4"	Each	493.00	3,816.00

0804.0 ENTRANCE DOORS, FRAMES & STORE FRONT CONSTRUCTION
Prices Include Hardware & Safety Plate Glazing

		UNIT	COST
.1	ALUMINUM		
.11	Stock		
	Door and Frame - 3' x 7'	Each	1,590.00
	6' x 7'	Each	2,915.00
	Window or Store Front Framing - 10'	SqFt	29.00
.12	Custom		
	Door and Frame - 3' x 7'	Each	1,908.00
	6' x 7'	Each	3,657.00
	Window or Store Front Framing - 10'	SqFt	36.00
	Add for Anodized Finish		15%
.2	BRONZE		
	Door and Frame - 3' x 7'	Each	4,028.00
	Window or Store Front Framing	SqFt	70.00
.3	STAINLESS STEEL		
	Door and Frame - 3' x 7'	Each	3,604.00
	Window or Store Front Framing	SqFt	72.00
	Add for Insulated Glass	SqFt	7.00
	Add for Exit Hardware	Each	371.00

See 0803.10 for Revolving Doors
See 0803.0 for Automatic Door Opening Assemblies

0805.0 METAL WINDOWS (M) (CARPENTERS OR STEEL ERECTORS) (SINGLE GLAZING INCLUDED)

All Example Sizes Below: 2'4" x 4'6"

		UNIT	LABOR	MATERIAL	or	UNIT	LABOR	MATERIAL
.1	ALUMINUM WINDOWS							
	Casement and Awning	Each	55.00	265.00		SqFt	5.30	25.30
	Sliding or Horizo Rolling	Each	55.00	191.00		SqFt	5.30	19.00
	Double and Single-Hung or Vertical Sliding	Each	56.00	191.00		SqFt	5.40	19.00
	Projected	Each	52.00	244.00		SqFt	5.05	24.00
	Add for Screens					SqFt	.70	4.20
	Add for Storms					SqFt	1.50	6.90
	Add for Insulated Glass					SqFt	-	7.15
	Add for Bronzed Finish					SqFt	-	10%
.2	ALUMINUM SASH							
	Casement	Each	45.00	244.00		SqFt	4.20	24.40
	Sliding	Each	45.00	170.00		SqFt	4.20	17.00
	Single-Hung	Each	45.00	170.00		SqFt	4.20	17.00
	Projected	Each	45.00	265.00		SqFt	4.20	26.50
	Fixed	Each	45.00	138.00		SqFt	4.20	14.00
.3	STEEL WINDOWS							
	Double-Hung	Each	52.00	387.00		SqFt	5.05	38.00
	Projected	Each	52.00	323.00		SqFt	5.05	32.30
	Add for Insulated Glass					SqFt	-	6.10
.4	STEEL SASH							
	Casement	Each	45.00	276.00		SqFt	4.20	28.00
	Double-Hung	Each	45.00	366.00		SqFt	4.20	38.00
	Projected	Each	45.00	313.00		SqFt	4.20	33.00
	Fixed	Each	45.00	180.00		SqFt	4.20	19.00

0806.0 WOOD WINDOWS (M) (Carpenters)

All Units Assembled, Kiln Dry Pine, Glazed, Vinyl Clad, Weather-stripped, and No Interior Trim. Top Quality. Other Sizes Available. Most commonly used listed below. Deduct 10% for Medium Quality and 25% for Low Quality Woods.

		UNIT	LABOR	MATERIAL
.1	BASEMENT OR UTILITY			
	Prefinished 2'8" x 1'4"	Each	42.00	126.00
	2'8" x 1'8"	Each	50.00	134.00
	2'8" x 2'0"	Each	52.00	145.00
	Add for Double Glazing	Each	-	42.00
	Add for Storm	Each	16.40	30.00
	Add for Screen	Each	13.25	19.00
.2	CASEMENT OR AWNING - Operable Units			
	Single Clear Tempered Glass 1'8" x 3'0"	Each	54.00	315.00
	1'8" x 3'6"	Each	55.00	326.00
	2'4" x 3'6"	Each	63.00	347.00
	2'4" x 4'0"	Each	65.00	415.00
	2'4" x 5'0"	Each	72.00	462.00
	Double (2 Units) 2'10" x 3'0"	Each	78.00	525.00
	3'4" x 3'6"	Each	84.00	651.00
	4'0" x 3'0"	Each	78.00	546.00
	4'0" x 3'6"	Each	85.00	672.00
	4'0" x 5'0"	Each	89.00	809.00
	4'0" x 5'8"	Each	93.00	882.00
	Triple (3 Units, 1 Fixed) 6'0" x 3'0"	Each	90.00	798.00
	6'0" x 4'0"	Each	95.00	924.00
	6'0" x 4'6"	Each	106.00	1,071.00
	Add for Grille Patterns - Interior - Polycarbonate	Each	-	23.00
	Add for Grille Patterns - Exterior - Prefinished	Each	-	63.00
	Add for High Performance Glazing	Each	-	56.00
	Add for Screen - per Unit	Each	-	20.00
	Add for Extension Jambs	Each	-	29.00
	Add for Blinds	Each	-	95.00
	CASEMENT OR AWNING - Stationary & Fixed Units			
	Single 1'8" x 3'0"	Each	54.00	263.00
	1'8" x 3'6"	Each	55.00	294.00
	2'4" x 3'6"	Each	63.00	305.00
	2'4" x 4'0"	Each	66.00	378.00
	2'4" x 5'0"	Each	73.00	410.00
	2'4" x 6'0"	Each	77.00	473.00
	Picture 4'0" x 4'0"	Each	47.00	488.00
	4'0" x 4'6"	Each	51.00	520.00
	4'0" x 6'0"	Each	53.00	693.00
	6'0" x 4'0"	Each	54.00	683.00
	6'0" x 6'0"	Each	59.00	903.00
	Add for Grille Patterns - Wood - 2'4"	Each	24.00	89.00
	Add for Grille Patterns - Wood - 4'6"	Each	30.00	142.00
.3	DOUBLE HUNG (Insulating Glass with Screen)			
	2'6" x 3'0"	Each	45.00	336.00
	2'6" x 3'6"	Each	48.00	347.00
	2'6" x 4'2"	Each	51.00	368.00
	3'2" x 3'0"	Each	57.00	357.00
	3'2" x 3'6"	Each	59.00	389.00
	3'2" x 4'2"	Each	63.00	404.00
	Deduct for Primed Units	Each	5.30	28.00
	Add for Grille Patterns	Each	8.50	28.00
	Add for Triple Glazing	Each	-	66.00
	Deduct for Screen Unit	Each	10.60	21.00

0806.0 WOOD WINDOWS, Cont'd...

		UNIT	LABOR	MATERIAL
.4	GLIDER WITH SCREEN & INSUL GLASS 4'0" x 3'6"	Each	57.00	974.00
	5'0" x 3'6"	Each	66.00	971.00
	4'0" x 4'0"	Each	77.00	945.00
	5'0" x 4'0"	Each	83.00	1,024.00
.5	PICTURE WINDOW WITH CASEMENTS (2)			
	& INSULATING GLASS 9'6" x 5'0"	Each	148.00	1,890.00
	10'1" x 5'0"	Each	154.00	1,995.00
	9'6" x 5'6"	Each	148.00	1,995.00
	SINGLE GLASS AND STORM 10'3" x 4'6"	Each	133.00	2,021.00
	10'3" x 5'6"	Each	148.00	2,074.00
	10'3" x 6'6"	Each	154.00	2,153.00
	Add for High Performance Glass	Each	-	168.00
.6	CASEMENT ANGLE BAY WINDOWS, INSULATING GLASS, VINYL SHEATHED			
	30° (3 units) 5'10" x 4'2" x 2 7/8"	Each	207.00	1,523.00
	(4 units) 7'10" x 4'2" x 2 7/8"	Each	212.00	1,838.00
	(3 units) 5'10" x 5'2" x 2 7/8"	Each	201.00	1,649.00
	(4 units) 7'10" x 5'2" x 2 7/8"	Each	212.00	1,838.00
	(5 units) 9'10" x 4'2" x 2 7/8"	Each	249.00	2,153.00
	(5 units) 9'10" x 5'2" x 2 7/8"	Each	260.00	2,363.00
	45° (3 units) 5'4" x 4'2" x 2 7/8"	Each	212.00	1,523.00
	(4 units) 7'4" x 4'2" x 2 7/8"	Each	210.00	1,659.00
	(3 units) 5'4" x 5'2" x 2 7/8"	Each	212.00	1,659.00
	(4 units) 7'4" x 5'2" x 2 7/8"	Each	223.00	1,838.00
	(5 units) 9'4" x 4'2" x 2 7/8"	Each	239.00	2,258.00
	(5 units) 9'4" x 5'2" x 2 7/8"	Each	260.00	2,468.00
	Add for Screens	Each	50.00	53.00
	Add for Combination Windows	Each	-	257.00
	Deduct for Triple Glazing	Each	-	268.00
	Add for High Performance Glazing	Each	-	71.00
	Add for Bronze Glazing	Each	-	155.00
	Add for Color	Each	-	72.00
	Add for Blinds	Each	-	378.00
	Add for Bottom or Roof Skirt - Plywood	Each	100.00	420.00
	Add for Bottom or Roof Skirt - Aluminum	Each	106.00	604.00
.7	CASEMENT BOW WINDOWS, INSULATING GLASS			
	(3 units) 6'2" x 4'2"	Each	313.00	1,313.00
	(4 units) 8'2" x 4'2"	Each	376.00	1,764.00
	(3 units) 6'2" x 5'2"	Each	323.00	1,638.00
	(4 units) 8'2" x 5'2"	Each	403.00	2,048.00
.8	90° CASEMENT BOX BAY WINDOWS, INSULATING GLASS			
	4'8" x 4'2"	Each	408.00	1,869.00
	6'8" x 4'2"	Each	496.00	2,363.00
	4'8" x 5'2"	Each	419.00	1,901.00
	6'8" x 5'2"	Each	530.00	2,468.00
	Add for Screens - per Unit	Each	50.00	72.00
	Deduct for Double Glazing - per Unit	Each	-	39.00
	Add for High Performance Glazing	Each	-	78.00

		UNIT	LABOR	STATIONARY	VENT
.9	ROOF WINDOWS - INSULATING GLASS				
	1'10" x 3'10"	Each	194.00	431.00	704.00
	2'4" x 3'10"	Each	201.00	525.00	782.00
	3'8" x 3'10"	Each	220.00	630.00	945.00
	Add for Shingle Flashing	Each	43.00	82.00	78.00
	Add for High Performance Glass	Each	-	87.00	84.00

		UNIT	LABOR	MATERIAL
.10	CIRCLE TOPS - 2'4"	Each	113.00	551.00
	4'0"	Each	124.00	714.00
	4'8"	Each	130.00	840.00
	6'0"	Each	142.00	1,365.00

0807.0	SPECIAL WINDOWS (M) (Carpenters/ Steel Workers)	UNIT	LABOR	MATERIAL
.1	LIGHT-PROOF WINDOWS	SqFt	5.70	32.00
.2	PASS WINDOWS	SqFt	4.90	24.00
.3	DETENTION WINDOWS			
	Aluminum	SqFt	5.10	35.00
	Steel	SqFt	5.20	33.00
.4	VENETIAN BLIND WINDOWS (ALUMINUM)	SqFt	4.80	27.00
.5	SOUND-CONTROL WINDOWS	Each	4.45	21.00
0808.0	**DOOR AND WINDOW ACCESSORIES**			
.1	STORMS AND SCREENS (CARPENTERS)			
	Windows			
	Screen Only - Wood - 3' x 5'	Each	17.80	82.00
	Aluminum - 3' x 5'	Each	18.90	93.00
	Storm & Screen Combination - Aluminum	Each	21.00	113.00
	Doors			
	Screen Only - Wood - 3' x 6' - 8'	Each	31.00	200.00
	Aluminum	Each	29.00	210.00
	Storm & Screen Combination - Aluminum	Each	34.00	315.00
	Wood 1 1/8"	Each	36.00	300.00
.2	DETENTION SCREENS (M) (CARPENTERS)			
	4' 0" x 7' 0"	Each	89.00	551.00
.3	DOOR OPENING ASSEMBLIES (L&M) (GLAZIERS)			
	Floor or Overhead Electric Eye Units			
	Swing - Single 3' x 7' Door - Hydraulic	Each	-	4,200.00
	Double 6' x 7' Door - Hydraulic	Each	-	5,775.00
	Sliding - Single 3' x 7' Door - Hydraulic	Each	-	5,250.00
	Double 5' x 7' Door - Hydraulic	Each	-	7,350.00
	Industrial Door - 10' x 8	Each	-	7,140.00
.4	SHUTTERS - FOLDING (CARPENTERS)			
	16" x 1 1/8" x 48"	Pair	17.80	100.00
	x 60"	Pair	20.00	116.00
	x 72"	Pair	24.00	126.00

0809.0	CURTAIN WALL SYSTEMS (L&M) (IRONWORKERS)	UNIT	COST
.1	STRUCTURAL OR VERTICAL TYPE		
	Aluminum Tube Frame, Panel, Sills and Mullions - No glazing		
	Example: 4' Panel, 7' Fixed	SqFt	47.00 to 143.00
.2	PANEL WALL OR HORIZONTAL TYPE		
	Aluminum Tube Frame, Panel, Sills and Mullions - No glazing		
	Example: 1' Panel, 2' Oper., 4' Fixed	SqFt	42.00 to 117.00
.3	ARCHITECTURAL PANELS, INSULATED		
	Included in Above Prices		
	Metal		
	Baked Enamel on Steel - 24 ga	SqFt	15.50
	on Aluminum - 24 ga	SqFt	16.00
	Porcelain Enamel on Steel - 24 ga	SqFt	17.50
	on Aluminum - 24 ga	SqFt	17.90
	Exposed Aggregate - Resin	SqFt	13.60
	Fiberglass	SqFt	18.00
	Spandrel Glass	SqFt	16.00

See 0812 for Glazing Costs to be Added to Above

0810.0 FINISH HARDWARE (M) (CARPENTERS)

		EACH	LABOR	MATERIAL Painted	Bronze	Chrome
.1	BUTTS					
	3" x 3"	Pair	16.40	9.50	10.60	14.00
	3 1/2" x 3 1/2"	Pair	17.00	11.10	10.60	15.40
	4" x 4"	Pair	18.55	14.30	13.80	16.00
	4 1/2" x 4 1/2"	Pair	19.10	19.10	19.00	26.50
	4" x 4" Ball Bearing	Pair	20.10	36.00	36.00	45.60
	4 1/2" x 4 1/2" Ball Bearing	Pair	21.20	41.30	41.00	47.70

		UNIT	LABOR	MATERIAL
.2	CATCHES - ROLLER	Each	16.40	17.00
.3	CLOSURES - SURFACE MOUNTED - 3' - 0" Door	Each	50.00	111.00
	3' - 4"	Each	51.00	114.00
	3' - 8"	Each	52.00	117.00
	4' - 0"	Each	58.00	127.00
	CLOSURES - CONCEALED OVERHEAD - Interior	Each	84.00	175.00
	Exterior	Each	88.00	292.00
	Add for Fusible Link - Electric	Each	22.00	180.00
	CLOSURES - FLOOR HINGES - Interior	Each	122.00	557.00
	Exterior	Each	122.00	583.00
	Add for Hold Open Feature	Each	-	81.00
	Add for Double Acting Feature	Each	-	265.00
.4	DEAD BOLT LOCK - Cylinder - Outside Key	Each	39.00	138.00
	Cylinder - Double Key	Each	45.00	154.00
	Flush - Push/ Pull	Each	24.00	25.00
.5	EXIT DEVICES (PANIC) - Surface	Each	109.00	551.00
	Mortise Lock	Each	113.00	657.00
	Concealed	Each	159.00	2,014.00
	(Hand drop) Automatic	Each	124.00	1,484.00
.6	HINGES, SPRING (PAINTED) - 6" Single Acting	Each	30.00	69.00
	6" Double Acting	Each	30.00	88.00
.7	LATCHSETS - Bronze or Chrome	Each	40.00	143.00
	Stainless Steel	Each	40.00	164.00
.8	LOCKSETS - Mortise - Bronze or Chrome - H.D.	Each	37.00	212.00
	S.D.	Each	35.00	154.00
	Stainless Steel	Each	35.00	286.00
	Cylindrical - Bronze or Chrome	Each	39.00	170.00
	Stainless Steel	Each	39.00	217.00
.9	LEVER HANDICAP - Latch Set	Each	37.00	180.00
	Lock Set	Each	37.00	223.00
.10	PLATES - Kick - 8" x 34" - Aluminum	Each	22.00	22.00
	Bronze	Each	23.00	47.00
	Push - 6" x 15"- Aluminum	Each	21.00	23.00
	Bronze	Each	22.00	38.00
	Push & Pull Combination - Aluminum	Each	24.00	53.00
	Bronze	Each	24.00	88.00
.11	STOPS AND HOLDERS			
	Holder - Magnetic (No Electric)	Each	48.00	133.00
	Bumper	Each	28.00	20.00
	Overhead - Bronze, Chrome or Aluminum	Each	28.00	72.00
	Wall Stops	Each	22.00	16.00
	Floor Stops	Each	22.00	17.00
	See 0808.3 for Automatic Openers & Operators			

0811.0 WEATHERSTRIPPING (L&M) (CARPENTERS)

		UNIT	LABOR	MATERIAL
.1	ASTRAGALS - Aluminum - 1/8" x 2"	Each	22.00	15.00
	Painted Steel	Each	22.00	15.00
.2	DOORS (WOOD) - Interlocking	Each	37.00	34.00
	Spring Bronze	Each	32.00	55.00
	Add for Metal Doors	Each	32.00	10.00
.3	SWEEPS - 36" WOOD DOORS - Aluminum	Each	17.00	16.00
	Vinyl	Each	17.00	11.00
.4	THRESHOLDS - Aluminum - 4" x 1/2"	Each	21.00	18.00
	Bronze - 4" x 1/2"	Each	22.00	34.00
	5 1/2" x 1/2"	Each	23.00	39.00
.5	WINDOWS (WOOD) - Interlocking	Each	40.00	28.00
	Spring Bronze	Each	35.00	50.00

0812.0	**GLASS AND GLAZING (L&M) (Glaziers)**	UNIT	COST
.1	BEVELED GLASS - 1/4" x 1/2" Bevel	SqFt	127.00
.2	COATED - Used with Insulated Glass and Heat		
	Absorber and Light Reflector	SqFt	30.00
.3	HEAT ABSORBING AND SOLAR REJECTING COMBINATIONS		
	of Tinted, Coated and Clear Glass in Insulated Glass Form		
	with 1/4" or 1/2" Air Space. Usual thickness with 1/4" air		
	space, 9/16", 11/16" and 1 3/16". With 1/2" air space,		
	13/16", 15/16" and 1 /16". Factory priced. Greatly variable.	SqFt	22.00
.4	INSULATED GLASS		
	1/2" and 5/8" Welded	SqFt	16.20
	1" Clear	SqFt	18.00
	Tempered Insulated	SqFt	29.00
	Triple Insulated	SqFt	25.00
	Add for Tinting	SqFt	10%
.5	LAMINATED (SAFETY AND SOUND)		
	7/32"	SqFt	15.40
	1/4"	SqFt	18.30
	3/8"	SqFt	19.50
	1/2"	SqFt	28.10
	1 3/16", 1 1/2", 1 3/4" and 2" Bulletproof	SqFt	58.30
.6	MIRROR GLASS		
	Copper Plated Back and Polished Edges	SqFt	14.20
	Unfinished	SqFt	8.75
.7	PATTERNED OR ROUGH (OBSCURE)		
	1/8"	SqFt	7.10
	7/32"	SqFt	7.30
.8	PLASTIC GLASS (SAFETY)		
	Acrylic - 1/8" .125"	SqFt	12.20
	3/16" .187"	SqFt	14.85
	1/4" .250"	SqFt	15.50
	5/16" .312"	SqFt	17.80
	3/8" .375"	SqFt	22.30
	Polycarbonates - Add to Above	SqFt	50%
.9	PLATE - CLEAR		
	1/4" (See .4 for Tinted)	SqFt	7.95
	3/8"	SqFt	10.90
	1/2"	SqFt	17.00
	5/8"	SqFt	19.90
	3/4"	SqFt	23.00
.10	SANDBLASTED - 3/16"	SqFt	10.10
.11	SHEET GLASS		
	Light Sheet or Window Glass - Double Strength	SqFt	4.90
	Heavy Sheet or Crystal	SqFt	6.70
.12	STAINED GLASS - Abstract	SqFt	88.00
	Symbolism	SqFt	127.00
	Figures	SqFt	297.00
.13	STRUCTURAL GLASS - 1/4"	SqFt	13.00
.14	TEMPERED (SAFETY PLATE)		
	1/4"	SqFt	9.10
	3/8"	SqFt	11.90
	1/2"	SqFt	25.30
	5/8"	SqFt	29.40
.15	TINTED PLATE - LIGHT REFLECTIVE		
	1/4"	SqFt	9.75
	3/8"	SqFt	13.35
	1/2"	SqFt	22.80
.16	WIRED (SAFETY)		
	1/4" - Clear	SqFt	14.70
	Obscure	SqFt	10.60

The costs below are average, priced as total contractor / subcontractor costs with an overhead and fee of 10% included. Units include fasteners, tools and equipment, fringe benefits, 30% taxes and insurance on labor, and 5% General Conditions.

0801.0 HOLLOW METAL (Installation of Butts and Locksets included but no hardware material)

	UNIT	COST
.1 CUSTOM FRAMES (16 ga)		
2'6" x 6'8" x 4 3/4"	Each	265
2'8" x 6'8" x 4 3/4"	Each	286
3'0" x 6'8" x 4 3/4"	Each	297
3'4" x 6'8" x 4 3/4"	Each	318
Add for A, B or C Label	Each	34
Add for Side Lights	Each	154
Add for Frames over 7'0" - Height	Each	64
Add for Frames over 6 3/4" - Width	Each	55
CUSTOM DOORS (1 3/8" or 1 3/4") (18 ga)		
2'6" x 6'8" x 1 3/4"	Each	318
2'8" x 6'8" x 1 3/4"	Each	329
3'0" x 6'8" x 1 3/4"	Each	350
3'4" x 6'8" x 1 3/4"	Each	355
Add for 16 ga	Each	64
Add for B and C Label	Each	35
Add for Vision Panels or Lights	Each	80
.2 STOCK FRAMES (16 ga)		
2'6" x 6'8" or 7'0" x 4 3/4"	Each	191
2'8" x 6'8" or 7'0" x 4 3/4"	Each	196
3'0" x 6'8" or 7'0" x 4 3/4"	Each	207
3'4" x 6'8" or 7'0" x 4 3/4"	Each	223
Add for A, B or C Label	Each	34
STOCK DOORS (1 3/8" or 1 3/4" - 18 ga)		
2'6" x 6'8" or 7'0"	Each	302
2'8" x 6'8" or 7'0"	Each	313
3'0" x 6'8" or 7'0"	Each	334
3'4" x 6'8" or 7'0"	Each	355
Add for B or C Label	Each	30

0802.0 WOOD DOORS (Installation of Butts and Locksets included but no hardware material)

	UNIT	Hollow Core	Solid Core
.1 FLUSH DOORS - No Label - Paint Grade			
Birch - 7 Ply 2'4" x 6'8" x 1 3/8"	Each	189	210
2'6" x 6'8" x 1 3/8"	Each	194	215
2'8" x 6'8" x 1 3/8"	Each	200	226
3'0" x 6'8" x 1 3/8"	Each	205	242
2'4" x 6'8" x 1 3/4"	Each	200	231
2'6" x 6'8" x 1 3/4"	Each	205	236
2'8" x 6'8" x 1 3/4"	Each	210	257
3'0" x 6'8" x 1 3/4"	Each	221	257
3'4" x 6'8" x 1 3/4"	Each	226	278
3'6" x 6'8" x 1 3/4"	Each	231	389
Add for Stain Grade	Each		36
Add for 5 Ply	Each		79
Add for Jamb & Trim - Solid - Knock Down	Each		81
Add for Jamb & Trim - Veneer - Knock Down	Each		53
Add for Red Oak - Rotary Cut	Each		32
Add for Red Oak - Plain Sliced	Each		42
Deduct for Lauan	Each		24
Add for Architectural Grade - 7 Ply	Each		79
Add for Vinyl Overlay	Each		58
Add for 7' - 0" Doors	Each		24
Add for Lite Cutouts with Metal Frame	Each		88
Add for Wood Louvres	Each		131
Add for Transom Panels & Side Panels	Each		168

0802.0 WOOD DOORS, Cont'd...

		UNIT	MATERIAL		
.1	FLUSH DOORS, Cont'd...		20-Min	45-Min	60-Min
	Label - 1 3/4" - Paint Grade				
	2'6" x 6'8" - Birch	Each	263	294	294
	2'8" x 6'8"	Each	273	305	315
	3'0" x 6'8"	Each	284	326	326
	3'4" x 6'8"	Each	305	352	352
	3'6" x 6'8"	Each	368	362	362

		UNIT	Pine	Fir	Birch/ Oak	Lauan
.2	PANEL DOORS					
	Exterior - 1 3/4"					
	2'8" x 6'8"	Each	546	557	777	-
	3'0" x 6'8"	Each	578	588	809	-
	Interior - 1 3/8"					
	2'6" x 6'8"	Each	473	420	609	-
	2'8" x 6'8"	Each	483	431	630	-
	3'0" x 6'8"	Each	515	441	662	-
	Add for Sidelights	Each	305	441	494	
.3	LOUVERED DOORS - 1 3/8"					
	2'0" x 6'8"	Each	320	315	662	-
	2'6" x 6'8"	Each	373	368	683	-
	2'8" x 6'8"	Each	383	378	693	-
	3'0" x 6'8"	Each	394	389	714	-
.4	BI-FOLD - 2 DOOR WITH HDWE - 1 3/8"					
	2'0" x 6'8" - Flush	Each	-	-	236	200
	2'4" x 6'8"	Each	-	-	247	210
	2'6" x 6'8"	Each	-	-	257	221
	Add for Prefinished	Each				24
	See 0804.22 for 1 1/8" Closet Bi-folds					
.5	CAFE DOORS - 1 1/8" - Pair					
	2'6" x 3'8"	Each	431	-	-	-
	2'8" x 3'8"	Each	441	-	-	-
	3'0" x 3'8"	Each	462	-	-	-
.6	FRENCH DOORS - 1 3/8"					
	2'6" x 3'8"	Each	725	578	777	-
	2'8" x 3'8"	Each	735	588	809	-
	3'0" x 3'8"	Each	756	599	819	-
.7	DUTCH DOORS					
	2'6" x 3'8"	Each	924	-	-	-
	2'8" x 3'8"	Each	987	-	-	-
	3'0" x 3'8"	Each	1,050	-	-	-
.8	PREHUNG DOOR UNITS (INCL. FRAME, TRIM, HARDWARE, and SPLIT JAMBS)					
	Exterior Entrance - 1 3/4"					
	Panel - 2'8" x 6'8"	Each	504	-	-	-
	3'0" x 6'8"	Each	536	-	-	-
	3'0" x 7'0"	Each	578	-	-	-
	Add for Insulation	Each	210			
	Add for Side Lights	Each	221			
	Interior - 1 3/8"					
	Flush, H.C. - 2'6" x 6'8"	Each	-	-	362	273
	2'8" x 6'8"	Each	-	-	373	278
	3'0" x 6'8"	Each	-	-	394	289
	Add for Int. Panel Door	Each				252

0803.0   SPECIAL DOORS

.2	CLOSET BI-FOLDING - PREHUNG			COST		
	STOCK AND UNFINISHED	UNIT	Oak Flush	Birch Flush	Lauan Flush	Pine Panel
	Wood - 2 Door - 2'0" x 6'8" x 1 3/8"	Each	286	329	244	541
	2'6" x 6'8" x 1 3/8"	Each	302	350	254	562
	3'0" x 6'8" x 1 3/8"	Each	323	403	265	636
	4 Door - 4'0" x 6'8" x 1 3/8"	Each	366	413	318	875
	5'0" x 6'8" x 1 3/8"	Each	376	435	371	975
	6'0" x 6'8" x 1 3/8"	Each	424	445	382	1,081

		UNIT	COST
	Add for Prefinished -	Each	$35.00
	Add for Jambs and Casings -	Each	$82.00
	Metal - 2 Door - 2'0" x 6'8"	Each	164
	2'6" x 6'8"	Each	170
	3'0" x 6'8"	Each	199
	4 Door - 4'0" x 6'8"	Each	228
	5'0" x 6'8"	Each	239
	6'0" x 6'8"	Each	265
	Add for Plastic Overlay	Each	40
	Add for Louvre or Decorative Type	Each	18
	Leaded Mirror (Based on 2-Panel Unit)		
	4'0" x 6'8"	Each	806
	5'0" x 6'8"	Each	859
	6'0" x 6'8"	Each	859
.3	FLEXIBLE DOORS (M) (CARPENTERS)		
	Wood Slat (Unfinished)	SqFt	1,696
	Fabric Accordion Fold	SqFt	1,590
	Vinyl Clad	SqFt	2,014
.8	PLASTIC LAMINATE FACED (M) (CARPENTERS) - 1 3/8"	SqFt	2,120
	1 3/4"	SqFt	2,120
	Add for Decorative Type	SqFt	2,226
	Add for Label Doors	SqFt	5.60
	Add for Edge Strip - Vertical and Horizontal	Each	10.00
	Add for Machining for Hardware	Each	53.00
.10	ROLLING - DOORS & GRILLES		
	Doors - 8' x 8'	Each	2,968
	10' x 10'	Each	3,180
	Grilles - 6'-8" x 3'-2"	Each	1,060
	6'-8" x 4'-2"	Each	1,166
.11	SHOWER DOORS - 28" x 66" (CARPENTERS)	Each	254
.12	SLIDING OR PATIO DOORS (M) (CARPENTERS)		
	Metal (Aluminum) - Including Glass Thresholds & Screen		
	Opening 8'0" x 6'8"	Each	1,722
	8'0" x 8'0"	Each	1,829
	Wood		
	Vinyl Clad 6'0" x 6'10"	Each	2,120
	8'0" x 6'10"	Each	2,438
	Pine - Prefinished 6'0" x 6'10"	Each	2,120
	8'0" x 6'10"	Each	2,438
	Add for Grilles	Each	350
	Add for Triple Glazing	Each	233
	Add for Screen	Each	159
.13	SOUND REDUCTION - Metal	Each	2,067
	Wood	Each	1,378
.14	TRAFFIC DOORS (CARPENTERS)		
	Electric Powered - Hollow Metal - 8' x 8' Opening	Each	5,936
	Wood - 8' x 8' Opening	Each	5,830
	Truck Impact - Metal Clad - 8' x 8' Opening	Each	3,604
	(Double Acting) - Rubber - 8' x 8' Opening	Each	2,915
	Add for Frames	Each	890
.16	VAULT DOORS - 6'6" x 2'2" with Frame - 2 hour	Each	3,604

0805.0 METAL WINDOWS (M) (CARPENTERS OR STEEL ERECTORS)
(SINGLE GLAZING INCLUDED)

All Example Sizes Below: 2'4" x 4'6"

		UNIT	COST	UNIT	COST
.1	ALUMINUM WINDOWS				
	Casement and Awning	Each	366.00	SqFt	34.80
	Sliding or Horizontal	Each	292.00	SqFt	27.70
	Double and Single-Hung or Vertical Sliding	Each	323.00	SqFt	30.75
	Projected	Each	366.00	SqFt	34.80
	Add for Screens	Each	53.00	SqFt	5.05
	Add for Storms	Each	95.00	SqFt	9.00
	Add for Insulated Glass	Each	80.00	SqFt	7.50
.2	ALUMINUM SASH				
	Casement	Each	345.00	SqFt	32.75
	Sliding	Each	292.00	SqFt	27.70
	Single-Hung	Each	292.00	SqFt	27.70
	Projected	Each	244.00	SqFt	23.20
	Fixed	Each	254.00	SqFt	24.20
.3	STEEL WINDOWS				
	Double-Hung	Each	530.00	SqFt	50.35
	Projected	Each	509.00	SqFt	48.30
.4	STEEL SASH				
	Casement	Each	408.00	SqFt	38.80
	Double-Hung	Each	466.00	SqFt	44.30
	Projected	Each	350.00	SqFt	33.20
	Fixed	Each	360.00	SqFt	34.25

0806.0 WOOD WINDOWS (M) (Carpenters) - All Units Assembled, Kiln Dry Pine, Glazed, Vinyl Clad, Weatherstripped, and Top Quality. Many Other Sizes Available.

		UNIT	COST
.1	BASEMENT OR UTILITY		
	Prefinished with Screen 2'8" x 1'4"	Each	210
	2'8" x 2'0"	Each	247
	Add for Double Glazing or Storm	Each	53
.2	CASEMENT OR AWNING		
	Operating Units - Insulating Glass		
	Single 2'4" x 4'0"	Each	473
	2'4" x 5'0"	Each	525
	Double with Screen 4'0" x 4'0" (2 units)	Each	756
	4'0" x 5'0" (2 units)	Each	982
	Triple with Screen 6'0" x 4'0" (3 units)	Each	1,208
	6'0" x 5'0" (3 units)	Each	1,313
	Fixed Units - Insulating Glass		
	Single 2'4" x 4'0"	Each	452
	2'4" x 5'0"	Each	525
	Picture 4'0" x 4'6"	Each	641
	4'0" x 6'0"	Each	809
.3	DOUBLE HUNG - & Insul. Glass with Screen - 2'6" x 3'6"	Each	452
	2'6" x 4'2"	Each	462
	3'2" x 3'6"	Each	473
	3'2" x 4'2"	Each	536
	Add for Triple Glazing	Each	67
.4	GLIDER - & Insul. Glass with Screen - 4'0" x 3'6"	Each	1,103
	5'0" x 4'0"	Each	1,313
.5	PICTURE WINDOWS with CASEMENTS - & Insul. Glass with Screen		
	9'6" x 4'10"	Each	2,310
	9'6" x 5'6"	Each	2,520

0806.0   WOOD WINDOWS Cont'd...

.6   CASEMENT ANGLE BAY WINDOWS-Insulating Glass with Screens

30° - 5'10" x 4'2" - (3 units)		Each	1,785
7'10" x 4'2" - (4 units)		Each	2,258
5'10" x 5'2" - (3 units)		Each	1,890
7'10" x 5'2" - (4 units)		Each	2,205
45° - 5'4" x 4'2" - (3 units)		Each	1,785
7'4" x 4'2" - (4 units)		Each	1,995
5'4" x 5'2" - (3 units)		Each	2,153
7'4" x 5'2" - (4 units)		Each	2,363

.7   CASEMENT BOW WINDOWS - Insulating Glass

6'2" x 4'2" - (3 units)		Each	1,785
8'2" x 4'2" - (4 units)		Each	2,520
6'2" x 5'2" - (3 units)		Each	2,048
8'2" x 5'2" - (4 units)		Each	2,625
Add for Triple Glazing - per unit		Each	68
Add for Bronze Glazing - per unit		Each	68
Deduct for Primed Only - per unit		Each	17
Add for Screens - per unit		Each	21

.8   90° CASEMENT BOX BAY WINDOWS - Insulating Glass

4'8" x 4'2"		Each	2,520
6'8" x 4'2"		Each	3,308
6'8" x 5'2"		Each	3,465

	UNIT	STATIONARY	MOVABLE
.9 SKY OR ROOF WINDOWS 1'10" x 3'10"	Each	693	987
2'4" x 3'10"	Each	809	1,155
3'8" x 3'10"	Each	956	1,365
.10 CIRCULAR TOPS & ROUNDS - 4'0"	Each	-	756
6'0"	Each	-	1,628

0807.0   SPECIAL WINDOWS (M) (CARPENTERS/ STEEL WORKERS)

.1 LIGHT-PROOF WINDOWS		SqFt	44
.2 PASS WINDOWS		SqFt	33
.3 DETENTION WINDOWS		SqFt	46
.4 VENETIAN BLIND WINDOWS (ALUMINUM)		SqFt	38
.5 SOUND-CONTROL WINDOWS		SqFt	37

0808.0   DOOR AND WINDOW ACCESSORIES

.1   STORMS AND SCREENS (CARPENTERS)

Windows

Screen Only - Wood - 3' x 5'		Each	116
Aluminum - 3' x 5'		Each	134
Storm & Screen Combination - Aluminum		Each	150

Doors

Screen Only - Wood		Each	278
Aluminum - 3' x 6' - 8'		Each	275
Storm & Screen Combination - Aluminum		Each	331
Wood 1 1/8"		Each	378

.2   DETENTION SCREENS (M) (CARPENTERS)

Example: 4' 0" x 7' 0"		Each	777

.3   DOOR OPENING ASSEMBLIES (L&M) (GLAZIERS)

Floor or Overhead Electric Eye Units

Swing - Single 3' x 7' Door - Hydraulic		Each	4,515
Double 6' x 7' Door - Hydraulic		Each	6,825
Sliding - Single 3' x 7' Door - Hydraulic		Each	5,775
Double 5' x 7' Doors- Hydraulic		Each	7,770
Industrial Doors - 10' x 8'		Each	7,875

.4   SHUTTERS (CARPENTERS)

16" x 1 1/8" x 48"		Pair	131
16" x 1 1/8" x 60"		Pair	147
16" x 1 1/8" x 72"		Pair	152

0810.0 FINISH HARDWARE (M) (CARPENTERS)

				COST	
.1	BUTTS	UNIT	Painted	Bronze	Chrome
	3" x 3"	Pair	37	39	42
	3 1/2" x 3 1/2"	Pair	38	40	45
	4" x 4"	Pair	41	42	47
	4 1/2" x 4 1/2"	Pair	45	48	56
	4" x 4" Ball Bearing	Pair	69	71	80
	4 1/2" x 4 1/2" Ball Bearing	Pair	74	80	85

		UNIT	COST
.2	CATCHES - ROLLER	Each	42
.3	CLOSURES - SURFACE MOUNTED - 3' - 0" Door	Each	212
	3' - 4"	Each	217
	3' - 8"	Each	223
	4' - 0"	Each	239
	CLOSURES - CONCEALED OVERHEAD - Interior	Each	339
	Exterior	Each	477
	Add for Fusible Link - Electric	Each	239
	CLOSURES - FLOOR HINGES - Interior	Each	816
	Exterior	Each	848
	Add for Hold Open Feature	Each	98
	Add for Double Acting Feature	Each	329
.4	DEAD BOLT LOCK - Cylinder - Outside Key	Each	212
	Cylinder - Double Key	Each	254
	Flush - Push/ Pull	Each	68
.5	EXIT DEVICES (PANIC) - Surface	Each	816
	Mortise Lock	Each	954
	Concealed	Each	2,332
	Handicap (ADA) Automatic	Each	1,696
.6	HINGES, SPRING (PAINTED) - 6" Single Acting	Each	122
	6" Double Acting	Each	154
.7	LATCHSETS -Bronze or Chrome	Each	212
	Stainless Steel	Each	265
.8	LOCKSETS - Mortise - Bronze or Chrome - H.D.	Each	276
	Mortise - Bronze or Chrome - S.D.	Each	239
	Stainless Steel	Each	387
	Cylindrical - Bronze or Chrome	Each	249
	Stainless Steel	Each	318
.9	LEVER HANDICAP - Latch Set	Each	265
	Lock Set	Each	329
.10	PLATES - Kick - 8" x 34" - Aluminum	Each	98
	Bronze	Each	86
	Push - 6" x 15" - Aluminum	Each	55
	Bronze	Each	82
	Push & Pull Combination - Aluminum	Each	99
	Bronze	Each	143
.11	STOPS AND HOLDERS		
	Holder - Magnetic (No Electric)	Each	228
	Bumper	Each	55
	Overhead - Bronze, Chrome or Aluminum	Each	117
	Wall Stops	Each	37
	Floor Stops	Each	49
	See 0808.3 for Automatic Openers & Operators		

0811.0 WEATHERSTRIPPING (L&M) (CARPENTERS)

		UNIT	COST
.1	ASTRAGALS - Aluminum - 1/8" x 2"	Each	49
	Painted Steel	Each	49
.2	DOORS (WOOD) - Interlocking	Each	91
	Spring Bronze	Each	104
	Add for Metal Doors	Each	49
.3	SWEEPS - 36" WOOD DOORS - Aluminum	Each	42
	Vinyl	Each	32
.4	THRESHOLDS - Aluminum - 4" x 1/2"	Each	49
	Bronze 4" x 1/2"	Each	74
	5 1/2" x 1/2"	Each	81
.5	WINDOWS (WOOD) - Interlocking	Each	83
	Spring Bronze	Each	104

All items in this division are subcontractors' costs and include all labor, materials, equipment and fees.

0901.0 LATH & PLASTER (L&M) (Latherers and Plaster)

		UNIT	COST
.1	LATHING		
.11	Metal Lath	SqYd	16.20
.12	Gypsum Lath	SqYd	14.00
.13	Metal Studs	SqYd	17.80
.14	Channels - Studs - 16" O.C.	SqYd	17.30
	Furring - Beams & Columns - 16" O.C.	SqYd	21.00
	Ceilings (including hangers) - 16" O.C.	SqYd	24.00
.15	Metal Trim - Base Bead and Corner Bead	LnFt	4.00
	Picture Mould	LnFt	5.30
.2	PLASTERING		
.21	Gypsum (Putty Coat, Sand Float Textured),Walls - 1 Coat	SqYd	17.85
	3 Coat	SqYd	50.00
	Add for Ceiling, Column and Beam Applications	SqYd	4.75
.22	Acoustical	SqYd	29.50
.23	Vermiculite	SqYd	26.20
.24	Stucco or Cement Plaster		
	On Masonry - Large Areas - Float Finish - 1 Coat	SqYd	30.00
	3 Coat	SqYd	38.80
	Panels, Facias, Soffits - 3 Coat	SqYd	46.00
	On Metal Lath - Large Areas - 3 Coat	SqYd	57.00
	Panels, Facias, Soffits, Columns	SqYd	62.00
	On Metal Studs - 4" - 16 ga	SqYd	36.50
	6" - 16 ga	SqYd	52.00
	Deduct for 18 ga Studs	SqYd	10%
	Add for Trowel Finish	SqYd	4.85
.25	Fireproofing (Sprayed On) - Mineral Fiber		
	Steel and Decks - 1 Hour	SqYd	11.00
	2 Hour	SqYd	18.60
	Beams and Columns - Not Wrapped - 2 Hour	LnFt	19.50
	3 Hour	LnFt	22.50
.26	Thin Coat or Veneer Plaster - Walls - 1 Coat	SqYd	15.75
	2 Coat	SqYd	24.20
	Ceilings - 2 Coat	SqYd	30.50
	Add for Patching	SqYd	50%
	Add Color to Plaster and Stucco	SqYd	5.50
	Add for Scaffold - Average	SqYd	9.60
.27	Exterior Wall System - Stucco or Synthetic Plaster		
	Includes Structural Board and 2" Insulation	SqYd	111.00
	Add for Esthetic Grooving	LnFt	5.30

LATH AND GYPSUM PLASTER IN COMBINATION UNITS:		FINISHED	
WALLS	UNIT	1 Side	2 Side
1 Coat Plaster on Waterproofed Found. Walls	SqYd	18.00	-
2 Coat Plaster on Masonry and Concrete	SqYd	35.50	71.00
on Gypsum Lath	SqYd	44.50	101.00
3 Coat Plaster on Metal Lath	SqYd	55.20	112.00
2 Coat Plaster on Metal Studs & Gypsum Lath	SqYd	73.50	120.00
CEILINGS			
3 Coat Plaster on Metal Lath and Channel for Suspended Ceiling (including hangers)	SqYd	105.00	-
3 Coat Plaster and Metal Lath attached to joists, beams, etc.	SqYd	88.00	-
FURRED			
Beam - 3 Coat Plaster, Metal Lath, Channel	SqYd	107.00	-
Column & Pipe - 3 Coat Plaster - Metal Lath	SqYd	100.00	-
Cabinet - 3 Coat Plaster, Metal Lath, Chnl.	SqYd	68.00	-
2 Coat Plaster on Gypsum Lath	SqYd	62.00	-

0902.0 GYPSUM DRYWALL (L&M) (Carpenters)
Based on 8' Ceiling Heights & Screwed-on Application

		UNIT	COST
.1	BACKER BOARD		
	3/8"	SqFt	1.10
	1/2"	SqFt	1.13
	5/8" Fire Rated	SqFt	1.14
	Deduct for Nailed-on	SqFt	.08
.2	FINISH BOARD - Walls		
	1/2" - Standard	SqFt	1.75
	Fire Rated	SqFt	1.77
	Moisture Resistant	SqFt	1.86
	Insulating (Foil Backed)	SqFt	1.91
	Combined Fire Rated and Insulating	SqFt	1.95
	5/8" - Standard	SqFt	1.74
	Fire Rated	SqFt	1.76
	Moisture Resistant	SqFt	1.76
	Insulating (Foil Backed)	SqFt	1.80
	Combined Fire Rated and Insulating	SqFt	1.90
	Lead Lined with #4 Lead - 1/16"	SqFt	8.50
	#2 Lead - 1/16"	SqFt	6.50
	Add for Ceiling Work - to Wood	SqFt	.18
	Add for Ceiling Work - to Steel	SqFt	.42
	Add for Adhesive Method	SqFt	.18
	Deduct for Nailed On	SqFt	.09
	Add for Beam and Column Work	SqFt	.78
	Add for Resilient Clip Application	SqFt	.90
	Add for Small Cut Up Areas	SqFt	.65
	Add for Each Floor Added	SqFt	.08
	Add for Work Above 10'	SqFt	.25
	Add for Filling Hollow Metal Frames	Each	85.00
.3	VINYL COATED BOARD - 1/2"	SqFt	2.15
	5/8"	SqFt	2.25
.4	LINER PANELS - 1"	SqFt	2.90

			COST	
		UNIT	25 ga	20 ga
.5	STEEL STUDS (Incl. Top & Bottom Plates)			
	16" O.C. - 1 5/8"	SqFt	1.50	1.70
	2 1/2"	SqFt	1.52	1.72
	3 5/8"	SqFt	1.48	1.68
	6"	SqFt	1.54	1.72
	24" O.C. - 1 5/8"	SqFt	1.16	1.38
	2 1/2"	SqFt	1.20	1.43
	3 5/8"	SqFt	1.21	1.48
	6"	SqFt	1.28	1.57
	Add for Work Over 10' Heights	SqFt	.18	.18
	Add for 16 Ga 6" Galvanized	SqFt	1.75	1.30
.6	FURRING CHANNELS			
	16" O.C. 3/4" x 1 3/8" walls	SqFt	-	1.24
	24" O.C. 3/4" x 1 3/8" walls	SqFt	-	1.14
	Resilient Channels - 16" O.C.	SqFt	-	1.14
	Add for Ceiling Work	SqFt	-	.16
	Add for Beam and Column Work	SqFt	-	1.20
.7	METAL TRIM - Casing Bead	LnFt	-	1.20
	Corner Bead	LnFt		.93
.8	TAPING AND SANDING	SqFt	-	.55
.9	TEXTURING	SqFt	-	.63

0902.0 GYPSUM DRYWALL, Cont'd...
COMBINATION UNITS - METAL FRAMING & BOARD

PARTITIONS 3 5/8" - 25 ga Studs - 24" O.C.	UNIT	COST
1/2" Board One Side	SqFt	3.05
Two Sides	SqFt	4.25
1/2" Fire Rated Board Two Sides	SqFt	4.30
1/2" Fire Rated Board One Side - 2 Layers 2nd Side (1 Hr.)	SqFt	5.40
2 Layers Each Side (2 Hr.)	SqFt	6.35
5/8" Board One Side	SqFt	3.10
Two Sides	SqFt	4.35
5/8" Fire Rated Board Each Side (1 Hr.)	SqFt	4.40
5/8" Fire Rated Board - 2 Layers Each Side (2 Hr.)	SqFt	6.65
3 Layers to Columns & Beams (2 Hr.)	SqFt	9.50
Add for Studs 16" O.C. - 25 Ga	SqFt	.29
Deduct for 1 5/8" Studs - 25 Ga	SqFt	.26
Deduct for 2 5/8" Studs - 25 Ga	SqFt	.14
Add for Studs - 20 Ga and 16" O.C.	SqFt	.32
Add for Studs - 20 Ga and 24" O.C.	SqFt	.30
Add for 6" Studs - 25 Ga and 16" O.C.	SqFt	.14
Add for 6" Studs - 20 Ga and 16" O.C.	SqFt	.24
Add for Sound Deadening	SqFt	.90
PARTITIONS - CAVITY SHAFT WALL -		
25 Ga Studs - 24" O.C. - All Board Fire Code C		
1" Liner Board One Side - 5/8" Board 2nd Side (1 Hr.)	SqFt	5.65
2 Layers - 1/2" Board 2nd Side (2 Hr.)	SqFt	5.20
3 Layers - 5/8" Board 2nd Side (3 Hr.)	SqFt	7.40
1" Liner Board and 5/8" Board One Side		
2 Layers 1" Liner Board 2nd Side	SqFt	9.90
PARTITIONS - DEMOUNTABLE		
2 1/2" Studs 24" O.C.-1/2" Vinyl Coated Bd. 2 Sides	SqFt	5.20
3 5/8" Studs 24" O.C.-1/2" Vinyl Coated Bd. 2 Sides	SqFt	5.60
Add for 16" O.C.	SqFt	.35
Add for 5/8" Vinyl Coated Board 2 Sides	SqFt	.38
Add for Special Colors Baked On	SqFt	25%
FURRED WALLS		
16" O.C. 1/2" Board	SqFt	2.83
24" O.C. 1/2" Board	SqFt	2.90
Resilient Channels and 1/2" Board	SqFt	2.85
CEILINGS		
To Wood Joists - 1/2" Fire Rated Board (1 Hr.)	SqFt	1.40
To Steel Joists with Furring Channels		
24" O.C. - 1/2" F.R. Board (2 Hr. with 2 1/2" Conc.)	SqFt	2.70
5/8" F.R. Board (1 Hr. with 2" Conc.)	SqFt	2.75
12" O.C. - 5/8" F.R. Board (2 Hr. with 2 1/2" Conc.)	SqFt	3.35
COLUMNS - 14 WF and Up		
1/2" F.R. Board with Screw Studs at Corners (1 Hr.)	SqFt	3.50
2 Layers - 1/2" F.R. Bd. with Screw Studs @ Corners (2 Hr.)	SqFt	4.00
Add for Corner Bead	LnFt	1.25

See 0703.0 for Insulation Work

		UNIT	COST
0903.0	**TILEWORK (L&M) (Tile Setters)**		
.1	CERAMIC TILE		
	Floors - 1" Square - Thin Set (Unglazed)	SqFt	8.80
	- 1" Hex - Thin Set	SqFt	9.50
	1" x 2"	SqFt	9.60
	2" Square	SqFt	9.00
	16" Square	SqFt	9.60
	Add for Mortar Setting Bed	SqFt	5.10
	Add for Random Colors	SqFt	1.20
	Add for Epoxy Joints	SqFt	2.60
	Conductive - In Mortar Setting Bed	SqFt	13.60
	Walls - 2" x 2" - Thin Set	SqFt	10.80
	4 1/4" x 4 1/4"	SqFt	8.20
	4 1/4" x 6"	SqFt	8.30
	4 1/4" x 8"	SqFt	8.60
	6" x 6"	SqFt	9.15
	Add for Plaster Setting Bed	SqFt	4.95
	Add for Decorator Type	SqFt	7.10
	Base - 4 1/4"	LnFt	8.50
	6"	LnFt	8.55
	Add for Coved Base	LnFt	1.00
	Accessories - Toilet and Bath Embedded	Each	63.00
.2	QUARRY TILE		
	Floors - 4" x 4" - Thin Set	SqFt	8.90
	6" x 6"	SqFt	8.80
	4" x 8"	SqFt	8.80
	Hexagon, Valencas, etc.	SqFt	10.00
	Walls - 4" x 4" - Thin Set	SqFt	9.60
	6" x 6"	SqFt	9.40
	Stairs	SqFt	12.40
	Base - 4" x 4" x 1/2"	LnFt	8.90
	6" x 6" x 1/2"	LnFt	10.20
	Add for Hydroment Joints	SqFt	.96
	Add for Epoxy Joints	SqFt	2.60
	Add for Non-Slip Tile	SqFt	.56
	Add for Mortar Setting Bed	SqFt	5.00
0904.0	**TERRAZZO (L&M) (Terrazzo Workers) - Based on Gray Cement, 4' Panel, and White Dividing Strips**		
.1	CEMENT TERRAZZO		
	Floors - 2" Bonded	SqFt	13.60
	2 1/2" Not Bonded	SqFt	11.30
	Base - 6" without Grounds or Base Bead	LnFt	20.30
	6" with Grounds and Base Bead	LnFt	26.00
	Stairs - Treads and Risers - Tread Length	LnFt	52.00
	Stringers without Base Bead	LnFt	47.00
	Pan Filled Treads and Landings	LnFt	22.00
	Add for White Cement	SqFt	.78
	Add for Brass Dividing Strips	SqFt	1.15
	Add for Large Chips (Venetian) - 3"	SqFt	6.25
	Add for 2-Color Panels	SqFt	3.70
	Add for 3-Color Panels	SqFt	4.40
	Add for Non-Slip Abrasives	SqFt	1.05
	Add for Small or Scattered Areas	SqFt	15%
.2	CONDUCTIVE TERRAZZO - 2" Bonded	SqFt	14.50
	1/4" Thin Set Epoxy	SqFt	12.75
.3	EPOXY TERRAZZO - 1/4" Thin Set	SqFt	10.50
.4	RUSTIC TERRAZZO (WASHED AGGREGATE) - Marble	SqFt	7.30
	Quartz	SqFt	7.90

See 0907 for Precast Terrazzo Tiles

0905.0 ACOUSTICAL TREATMENT (L&M) (Carpenters)

		UNIT	COST
.1	**MINERAL TILE (No Backing or Supports)**		
	Fissured or Perforated - 1/2" x 12" x 12" or 12" x 24"	SqFt	2.30
	5/8" x 12" x 12" or 12" x 24"	SqFt	2.60
	3/4" x 12" x 12" or 12" x 24"	SqFt	2.70
	Add to all above for Backing or Supports if needed:		
	Ceramic Faces	SqFt	.54
	Concealed Z Spline Application	SqFt	.68
	Wood Furring Strips	SqFt	.76
	Rock Lath	SqFt	1.00
	1 1/2" Channel Suspension	SqFt	1.40
	Resilient Furring Channels	SqFt	1.00
.2	**PLASTIC COVERED TILE (No Backing or Supports)**		
	3/4" x 12" x 12"	SqFt	2.55
	5/8" x 12" x 12"	SqFt	2.15
.3	**METAL PAN TILE (Supports Included)**		
	12" x 24" or 12" x 36" or 12" x 48" - Steel (24 Ga)	SqFt	7.80
	Aluminum (.025)	SqFt	9.60
	Aluminum (.032)	SqFt	11.00
	Stainless Steel	SqFt	13.20
	Add for Color	SqFt	.55
.4	**CLASS A MINERAL TILE**		
	(Grid System with Suspension incl. & Partitions to Clg.)		
	5/8" x 24" x 24" or 24" x 48" - Perforated	SqFt	1.85
	Fissured	SqFt	1.86
	Add for Partitions thru Ceilings	SqFt	.28
	Add for Translucent Panels for Lights	SqFt	1.55
.5	**FIRE RATED DESIGN MINERAL TILE**		
	(Grid System with Suspension incl. & Partitions to Clg.)		
	5/8" x 24" x 24" or 24" x 48" - Perforated (2-hour rating)	SqFt	2.02
	Fissured (2-hour rating)	SqFt	2.04
	3/4" x 24" x 24" or 24" x 48" - Perforated (3-4 Hr rating)	SqFt	2.20
	Fissured (3-4 Hr rating)	SqFt	2.21
	Add for Partitions thru Ceilings	SqFt	.28
	Add for Reveal Edge or Shadow Line (24" x 24" only)	SqFt	.38
	Add for Nubby Tile	SqFt	1.40
	Add for Ceramic Faces (5/8")	SqFt	.85
	Add for Colored Tile	SqFt	.29
	Add for Fixture Fire Protection - Fixtures only	SqFt	1.18
.6	**GLASS FIBRE BOARD (Grid System with Suspension included)**		
	5/8" x 24" x 24"	SqFt	2.00
	3/4" x 24" x 48"	SqFt	2.05
	1" x 48" x 48"	SqFt	2.00
	1-1/2" x 60" x 60"	SqFt	3.50
	Add for Three-Dimensional	SqFt	3.40
.7	**SOUND CONTROL PANELS AND BAFFLES**	SqFt	13.60

		UNIT	COST		
.8	**ACOUSTICAL WALL PANELS - Special**				
	Fiberglass with Cloth Face - 3/4"	SqFt	4.50	to	11.30
	1"	SqFt	5.70	to	13.00
	1-1/2"	SqFt	8.90	to	15.10
	3/4" Mineral	SqFt	1.45	to	4.00
	1/2" Fibre Board	SqFt	.95	to	1.40
.9	**LINEAR METAL CEILINGS**				
	4" Wide - Painted	SqFt	6.50	to	8.50
	Chrome or Brass	SqFt	9.40	to	12.00
	See 0307 for Fibre Cementitious Acoustical Plank				

0906.0 VENEER STONE (Floors, Walls, Stools, Stairs, Partitions)
(2" and Under) (L&M) (Marble Setters)

		UNIT	COST
.1	LIMESTONE		
.11	Cut Stone - Floors 7/8" Interior	SqFt	33.00
	Floors 2" Exterior	SqFt	37.20
	Walls 1"	SqFt	36.10
	Stools 1" x 8"	LnFt	43.50
.12	Flagstone Floors 1"	SqFt	40.00
.2	MARBLE Floors 5/8"	SqFt	34.40
	Floor Tiles 3/8" x 12" x 12"	SqFt	36.50
	Walls 3/8" x 12" x 12"	SqFt	40.20
	Base 1" x 6"	LnFt	29.75
	Stool 1" x 8"	LnFt	41.00
	Stairs - Treaded Risers	SqFt	79.00
.3	SLATE Floors 3/8" Package	SqFt	22.60
	Floors 3/4" Random or Flagstone	SqFt	23.60
	Walls 3/4" Random	SqFt	29.30
	Stools 1" x 8"	LnFt	42.00
.4	SANDSTONE Floors 2"	SqFt	37.80
.5	GRANITE Floors 5/6" Interior	SqFt	43.20
	Walls - 2" Exterior	SqFt	53.60
	1 1/2" Interior	SqFt	48.80
	Counter Tops 1 1/4"	SqFt	72.00
.6	SYNTHETIC MARBLE Floors 5/8"	SqFt	19.75
	Walls 5/8"	SqFt	22.70
	Add for Leveling Old Floors, per Square Foot		2.70

0907.0 SPECIAL FLOORING (L&M)

		UNIT	COST
.1	MAGNESIUM OXYCHLORIDE (Seamless)	SqFt	9.50
.2	ELASTOMERIC	SqFt	11.00
.3	SYNTHETIC		
	Epoxy with Granules 1/16" Single Speed	SqFt	5.50
	3/32" Double Speed	SqFt	6.60
	Epoxy with Aggregates 1/4" Industrial	SqFt	8.70
	Polyester with Granules 3/32"	SqFt	7.25
	1/4"	SqFt	9.00
	See 0904.3 for Epoxy Terrazzo and 0603.2 for Counter Tops		
.4	ASPHALTIC MASTIC	SqFt	8.70
	Add for Acid Proof	SqFt	1.60
.5	BRICK PAVERS - Mortar Setting Bed Application		
	3 7/8" x 8" x 1 1/4" - Red	SqFt	10.50
	Iron Spot	SqFt	12.60
	Acid Proof	SqFt	22.70
	See 0206.3 for Patios and Walks on Sand Cushion		
.6	PRECAST PANELS		
	12" x 12" x 1" Precast Terrazzo	SqFt	28.50
	12" x 12" x 1" Precast Exposed Aggregate	SqFt	28.70
	12" x 12" x 1" Ceramic (Terra Cotta)	SqFt	34.70
	6" Precast Terrazzo Base	LnFt	22.40
.7	NEOPRENE 1/4"	SqFt	12.50

0908.0 RESILIENT FLOORING (L&M)
(Carpet and Resilient Floor Layers)

		UNIT	COST
.1	ASPHALT - Tile 9" x 9" x 1/8"	SqFt	3.10
.2	CORK - Tile 1/8" - Many Sizes	SqFt	4.60
	3/16" - Many Sizes	SqFt	4.90
.3	RUBBER		
	Tile 12" x 12" x 3/32"	SqFt	5.70
	12" x 12" x 3/32" Conductive	SqFt	6.70
	Base 2 1/2"	LnFt	1.65
	4"	LnFt	1.90
	6"	LnFt	2.10
	Treads, Stringers and Risers		
	Treads 3/16" Gauge Pre-formed	LnFt	13.50
	1/4" Gauge Pre-formed	LnFt	15.50
	5/16" Gauge Pre-formed	LnFt	20.00
	Stringers 10" x 1/8"	LnFt	10.40
	Risers	LnFt	5.70
.4	VINYL COMPOSITION TILE 12" x 12" x 3/32"	SqFt	1.65
	12" x 12" x 1/8"	SqFt	1.80
	Conductive	SqFt	6.25
	Treads 1/8"	LnFt	6.60
	3/16"	LnFt	8.30
	1/4"	LnFt	10.40
	Stringers 10" x 1/8"	LnFt	10.50
	Risers 7" x 1/8"	LnFt	4.70
	Add for Marbleized or Patterns	SqFt	1.45
.5	VINYL TILE 9" x 9" x 1/8" Solid	SqFt	7.05
	12" x 12" x 1/8"	SqFt	8.20
	Base 2 1/2"	LnFt	1.70
	4"	LnFt	1.75
	6"	LnFt	1.00
.6	VINYL SHEET 1/8" Inlaid	SqFt	3.00 – 7.60
	1/8" Homogenous	SqFt	4.60 – 8.30
	Acoustic Cushioned	SqFt	5.00 – 12.00
	Coving	LnFt	6.65
.7	VINYL CORNER GUARDS	LnFt	6.00
	Add Plywood or Masonite Underlayment	SqFt	1.75
	Add Diagonal Cuttings	SqFt	.60
	Add Clean and Wax	SqFt	.45
	Add Skim Coat Leveling	SqFt	2.10 – 3.85

0909.0 CARPETING (L&M) (Carpet Layers)

		UNIT	CUT PILE	LOOP PILE
.1	NYLON - 20 oz	SqYd		15.20 - 26.50
	22 oz	SqYd		16.30 - 27.50
	24 oz	SqYd	13.60 – 20.00	17.50 - 32.00
	26 oz	SqYd		18.90 - 32.00
	28 oz	SqYd		19.00 - 34.00
	30 oz	SqYd	16.30 – 24.00	20.50 - 37.00
	32 oz	SqYd	17.60 – 25.00	22.00 - 39.00
.2	WOOL - 36 oz	SqYd		29.30 - 71.00
	42 oz	SqYd		37.60 - 84.00
.3	OLEFIN - 20 oz	SqYd		13.50 - 23.00
	22 oz	SqYd		15.00 - 24.00
	24 oz	SqYd		16.40 - 26.50
	26 oz	SqYd		17.60 - 29.50
	28 oz	SqYd		19.40 - 31.50
.4	CARPET TILE, MODULAR-Tension Bonded	SqYd		38.00 - 72.00
	Tufted	SqYd	26.00 – 40.00	26.00 - 52.00
.5	INDOOR-OUTDOOR	SqYd		10.00 - 24.00

Installation included in above prices

0909.0 CARPETING, Cont'd...

	UNIT	COST
Add to Section 0909.0 above:		
Add for Custom Carpet	SqYd	10.50 - 24.00
Add for Cushion (Urethane, Rubber, Synthetic)	SqYd	4.40 - 7.30
Add Bonded Cushion (Urethane Backed)	SqYd	1.50 - 2.00
Add for Carpet Base, Top Edge Bound	LnFt	2.30 - 3.65
Add for Skim Coat Leveling, Latex	SqFt	1.25 - 1.80
Add for Border and Inset Work	LnFt	2.70 - 4.70
Add for Reduction Strips	LnFt	2.00 - 3.15

0910.0 WOOD FLOORING (L&M) (Carpenters)

.1 STRIP AND PLANK FLOORING

Oak Floors (based on Standard Grade)	UNIT	COST
White Oak 25/32 1 1/2" Select Job Finished	SqFt	11.00
2 1/4" Select Job Finished	SqFt	11.20
3 1/4" Select Job Finished	SqFt	11.50
Add for Clear Grade	SqFt	.40
Deduct for Antique or Rustic Floors	SqFt	.50
Maple Floors - Iron Bound (Edge Grain)		
3/4" x 1 1/2" #2 & Better Gr Job Finished	SqFt	11.70
3/4" x 2 1/4" #2 & Better Gr Job Finished	SqFt	12.20
33/32 x 2 1/4" #2 & Better Gr Job Finished	SqFt	11.60
33/32 x 2 1/4" #3 Grade Job Finished	SqFt	10.90
Add for #1 Grade	SqFt	1.25
Add to above for Cork Underlayment	SqFt	1.35
Fir Floors		
1" x 4" Vertical Grain B&B Job Finished	SqFt	5.70
Flat Grain B&B Job Finished	SqFt	5.50

	UNIT	COST
.2 WOOD PARQUET Pre Finished 5/16" Oak or Maple	SqFt	8.90
FLOORING Pre Finished 3/4 " Oak or Maple	SqFt	11.00
Pre Finished 5/16" Teak	SqFt	12.10
5/16" Walnut	SqFt	12.40
.3 RADIANT TREATED (Plastic and Wood) 5/16" x 12" x 12"	SqFt	13.50
.4 RESILIENT WOOD FLOOR SYSTEMS		
Cushion Sleeper Floors (Maple)		
33/32 2 1/4" #1 Grade Pre-Finished	SqFt	9.90
#2 Grade Pre-Finished	SqFt	9.50
#3 Grade Pre-Finished	SqFt	8.80
Cork Underlayment included	SqFt	1.40
Add for Plywood Underlayment (2 layers)	SqFt	2.20
Fixed Sleeper Floors (Oak)		
33/32 2 1/4" #1 Grade Pre-Finished	SqFt	9.90
#2 Grade Pre-Finished	SqFt	9.80
1/4" #3 Grade Pre-Finished	SqFt	9.70
.5 WOOD BLOCK INDUSTRIAL FLOORS (Creosoted) 2"	SqFt	5.20
2 1/2"	SqFt	6.10
3"	SqFt	6.25
.6 FINISHING (Sanding, Sealing and Waxing)		
Included in Job Finished Prices Above for New House	SqFt	2.05
for New Gym	SqFt	2.00
for Old Gym	SqFt	3.35
Add to Section 0910.0 above:		
Small Cut-up Areas	SqFt	30%
Trim and Irregular Areas	SqFt	15%
Polyurethane Finish with Stain	SqFt	20%
Deduct for Residential Work	SqFt	10%

0911.0 SPECIAL COATINGS

		UNIT	COST		
.1	CEMENTITIOUS (Glazed Cement)	SqFt	2.80	-	4.60
.2	EPOXIES	SqFt	1.65	-	3.20
.3	POLYESTERS	SqFt	1.70	-	3.30
.4	VINYL CHLORIDE	SqFt	1.95	-	3.60
.5	AGGREGATE TO EPOXY - Blown on Quartz	SqFt	8.30		
	Troweled on Quartz	SqFt	8.60		

		UNIT	COST
.6	SPRAYED FIREPROOFING (with Perlite)		
	Column - 1" 2 Hour	LnFt	1.90
	1 3/8" 3 Hour	LnFt	2.20
	1 3/4" 4 Hour	LnFt	3.00
	Beams - 1 1/8" 2 Hour	LnFt	2.10
	1 1/4" 3 Hour	LnFt	2.20
	1 1/2" 4 Hour	LnFt	2.45
	Ceilings - 1/2" 1 Hour	SqFt	1.55
	5/8" 2 Hour	SqFt	1.60
	1" 3 Hour	SqFt	2.00

See 0901.25 for Other Fireproofing

0912.0 PAINTING (L&M) (Painters)

WALLS AND CEILINGS - BRUSH

	UNIT	COST
Brick 2 Coats	SqFt	.78
Concrete 2 Coats	SqFt	.70
Concrete Block - Interior 2 Coats	SqFt	.73
Exterior 2 Coats	SqFt	.75
Light Weight 2 Coats	SqFt	.78
2 Coats/Filler	SqFt	.96
Plaster 2 Coats	SqFt	.73
Sheet Rock 2 Coats	SqFt	.62
Steel Decking & Sidings 1 Coat	SqFt	.53
Stucco 2 Coats	SqFt	.89
Wood Siding & Shingles 2 Coats	SqFt	.87
Wood Paneling - Stain, Seal and Varnish	SqFt	1.60
Deduct for Roller Application	SqFt	.12
Deduct for Spray Application	SqFt	.17
Add for Scaffold Work	SqFt	.26
Add for Epoxy Paint	SqFt	.24

DOORS AND WINDOWS

	UNIT	COST
Door, Wood - 2 Coats	Each	46.00
Stain, Seal and Varnish	Each	67.00
Bi-fold Door - Stain, Seal and Varnish 48"	Each	68.00
60"	Each	76.00
72"	Each	91.00
Door Frame, Wood (incl. Trim) - 2 Coats	Each	42.00
Stain, Seal & Varnish	Each	52.00
Door, Metal (Factory Primed) - 1 Coat	Each	38.00
2 Coats	Each	50.00
Door Frame, Metal (Factory Primed) - 1 Coat	Each	34.00
2 Coats	Each	38.00
Window, Wood (including Frame & Trim) - 2 Coats	Each	51.00
Add for Storm Windows or Screens	Each	27.00
Add for Shutters - Pair	Pair	72.00
Window, Steel	Each	45.00
FLOORS, EPOXY	SqFt	2.20

	UNIT	COST
0912.0 PAINTING (L&M) (Painters)		
MISCELLANEOUS WOOD		
Cabinets - Enamel 2 Coats	SqFt	.95
Stain and Varnish	SqFt	1.15
Enamel 3 Coats	SqFt	1.40
Stain, Seal and Varnish	SqFt	1.80
Wood Trim - Stain, Seal and Varnish - Large	LnFt	1.80
Small or Decorative	LnFt	2.10
MISCELLANEOUS METAL - 1 Coat Sprayed		
Miscellaneous Iron and Small Steel	Ton	1.15
Structural Steel	Ton	.76
Structural Steel - 2 Sq.Ft. per Ln.Ft.	LnFt	.58
3 Sq.Ft. per Ln.Ft.	LnFt	.70
4 Sq.Ft. per Ln.Ft.	LnFt	1.00
5 Sq.Ft. per Ln.Ft.	LnFt	1.20
Add for Brush	LnFt	40%
Add for 2 Coats	LnFt	75%
Gutters, Down Spouts and Flashings - 4"	LnFt	1.55
5"	LnFt	1.60
6"	LnFt	1.80
Piping - Prime and 1 Coat - 4"	LnFt	.76
6"	LnFt	1.10
8"	LnFt	1.25
10"	LnFt	1.70
Radiators	SqFt	3.25
MISCELLANEOUS		
Gold and Silver Leaf	SqFt	48.00
Parking Lines - Striping	LnFt	.34
Wood Preservatives	SqFt	.58
Waterproof Paints - 2 Coats	SqFt	.98
Clear - 1 Coat	SqFt	.52
2 Coat	SqFt	.78
Taping and Sanding	SqFt	.52
Texturing	SqFt	.57

See 0910 for Wood Floor Finishing

		UNIT	COST		
0913.0	**WALL COVERINGS (L&M) (Painters)**				
.1	VINYL PLASTICS (54" Roll)				
	Light Weight	SqFt	1.60	-	2.10
	Medium Weight	SqFt	1.65	-	2.60
	Heavy Weight	SqFt	1.75	-	4.65
	Add for Sizing	SqFt	.22	-	.40
	Add for 27" and 36" Rolls	SqFt	.18	-	.42
.2	PLASTIC WALL TILE	SqFt	3.25	-	4.65
.3	WALLPAPER ($10.00-40.00 Roll, 36 S.F. to Roll)	SqFt	1.55	-	4.15
	Add or Deduct per Roll - $1.00	SqFt	.20	-	.82
.4	FABRICS (Canvas, etc.)	SqFt	1.65	-	3.65
.5	LEATHER	SqFt	2.25	-	3.80
.6	WOOD VENEERS	SqFt	5.70	-	10.80
.7	FIBERGLASS	SqFt	2.35	-	4.20
.8	CORK - 3/16"	SqFt	1.95	-	4.65
.9	GLASS - 5/16"	SqFt	1.30	-	2.60
.10	GRASS CLOTH	SqFt	1.85	-	3.40

1001.0 ACCESS PANELS (M)
(Lathers, Carpenters)

	UNIT	LABOR	Flush Flange	MATERIAL Drywall	UL Label
10" x 10"	Each	16.50	38.00	42.00	120.00
12" x 12"	Each	17.20	44.00	-	125.00
14" x 14"	Each	18.20	47.00	52.00	148.00
16" x 16"	Each	20.00	49.00	-	166.00
18" x 18"	Each	21.50	58.00	-	170.00
22" x 22"	Each	24.60	76.00	86.00	202.00
24" x 24"	Each	29.00	89.00	-	224.00
24" x 36"	Each	33.50	103.00	-	285.00
32" x 32"	Each	37.50	120.00	-	310.00
Add for Plaster Type					15%
Add for Locks					10.00

1002.0 CHALKBOARDS AND TACKBOARDS (Carpenters)

	UNIT	LABOR	MATERIAL
Chalkboards			
Slate - Natural 3/8" - New	SqFt	4.40	18.00
Resurfaced	SqFt	4.40	9.25
Porcelain Enameled Steel			
1/2" Foil Backed Gypsum and 24 Ga	SqFt	3.90	8.80
28 Ga	SqFt	3.90	8.30
Marker (White) Board	SqFt	3.90	12.20
Tackboard			
Cork - 1/4" on 1/4" Gypsum Backing - Washable	SqFt	1.86	7.25
Add for Colored Cork	SqFt	-	3.20
Deduct for Natural Style - Not Washable	SqFt	-	3.10
Trim (Aluminum)			
Trim-Slip & Snap (incl. Metal Grounds) 1/4#	LnFt	2.70	2.15
1/2#	LnFt	3.05	4.30
Map Rail - 3/4 lb/ft	LnFt	3.30	6.00
1/4 lb/ft	LnFt	3.05	3.40
Chalk Tray	LnFt	4.20	7.25
Average Trim Cost	SqFt	3.20	6.40
Combination Unit Costs (Average)			
Slate - 3/8" - with Tackboard and Trim	SqFt	4.35	15.50
Enameled Steel (24 Ga) with Tackbd. & Trim, 8' x 4'	SqFt	4.10	9.30
Marker (White) Board	SqFt	4.20	9.70
Tackboard with Trim	SqFt	4.30	12.30
Add Vertical and Horizontal Sliding Units	SqFt	-	6.00
Portable Chalkboard Tackboard - 4' x 6'	Each	69.00	970.00
Add for Revolving Type		-	330.00

1003.0 CHUTES (Laundry & Rubbish)(M)(Sheet Metal Workers)

	UNIT	LABOR	MATERIAL
Laundry or Linen			
(16 Ga Aluminized Steel and 10' Floor Height)			
1 Floor Opening - 24" Dia - per Opening	Opng	240.00	1,040.00
2 Floors or Openings	Opng	220.00	850.00
6 Floors or Openings	Opng	210.00	800.00
Average	LnFt	23.00	90.00
Rubbish			
(16 Ga Aluminized Steel)			
1 Floor Opening - 24" Dia - per Opening	Opng	235.00	1,030.00
2 Floors or Openings	Opng	220.00	880.00
6 Floors or Openings	Opng	210.00	770.00
Add for Explosion Vent - per Opening	Opng	-	520.00
Add for Stainless Steel	Opng	-	440.00
Add for 30" Diameter	Opng	22.00	205.00
Deduct for Galvanized	Opng	-	88.00

1004.0 COAT/ HAT RACKS & CLOSET ACCESS. (M) (Carpenters)

	UNIT	LABOR	MATERIAL
ALUMINUM	LnFt	14.00	82.00
STAINLESS STEEL	LnFt	14.00	102.00

1005.0 CUBICLE CURTAIN TRACK & CURTAINS (M) (Carpenters)

	UNIT	LABOR	MATERIAL
	LnFt	9.20	18.00

1006.0 DIRECTORIES & BULLETIN BOARDS (M) (Carpenters)
(Directories For Changeable Letters)

	UNIT	LABOR	MATERIAL
Interior - 48" x 36" Aluminum	Each	150.00	1,030.00
Stainless Steel	Each	155.00	1,180.00
Bronze	Each	155.00	1,640.00
60" x 36" Aluminum	Each	175.00	1,180.00
Stainless Steel	Each	180.00	1,750.00
Bronze	Each	180.00	2,620.00
Exterior - 48" x 36" Aluminum	Each	210.00	1,420.00
Stainless Steel	Each	220.00	2,060.00
Bronze	Each	220.00	2,700.00
60" x 36" Aluminum	Each	230.00	1,950.00
Stainless Steel	Each	230.00	2,600.00
Bronze	Each	230.00	3,520.00
Illuminated - 72" x 48" Aluminum	Each	230.00	6,150.00
Stainless Steel	Each	330.00	9,700.00
Bronze	Each	330.00	11,900.00

1007.0 DISPLAY & TROPHY CASES (M) (Carpenters)

	UNIT	LABOR	MATERIAL
Aluminum Frame 8' x 6' x 1'6"	Each	350.00	3,650.00
4' x 6' x 1'6"	Each	205.00	2,000.00

1008.0 FLOORS - ACCESS OR PEDESTAL (L&M) (Carpenters)
(Includes Laminate Topping)

	UNIT	LABOR	MATERIAL
Aluminum	SqFt		21.50
Steel - Galvanized	SqFt		14.50
Plywood - Metal Covered	SqFt		13.40
Add for Carpeting	SqFt		3.00

1009.0 FIREPLACE AND STOVES

1010.0 FIRE FIGHTING DEVICES (M) (Carpenters)

	UNIT	LABOR	MATERIAL
Extinguishers - 2½ Gal - Water Pressure	Each	26.00	85.00
Carbon Dioxide - 5 Lb	Each	26.00	200.00
10 Lb	Each	28.00	280.00
5 Lb	Each	27.00	62.00
Dry Chemical - 10 Lb	Each	27.00	76.00
20 Lb	Each	29.00	125.00
Halon - 2 1/2 Lb	Each	26.00	136.00
5 Lb	Each	22.00	163.00
Cabinets - Enameled Steel - 18 Ga			
12" x 27" x 8" - Recessed - Portable Ext	Each	47.00	76.00
24" x 40" x 8" - Recessed - Hose Rack with Nozzle	Each	90.00	157.00
Deduct for Surface Plastic	Each	-	22.00
Add for Aluminum	Each	-	69.00
Blanket & Cabinet - Enameled and Roll Type	Each	47.00	86.00

1011.0 FLAG POLES (M) (Carpenters)

	UNIT	LABOR	MATERIAL
Aluminum Tapered - Ground 20' - 5" Butt		195.00	1,130.00
25' - 5 1/2" Butt		205.00	1,590.00
30' - 6" Butt		265.00	1,730.00
35' - 7" Butt		355.00	2,200.00
40' - 8" Butt		480.00	3,600.00
50' - 10" Butt		600.00	4,950.00
60' - 12" Butt		790.00	8,500.00
Wall 15' - 4" Butt		200.00	100.00
Add for Architectural Grade			25%
Fiberglass Tapered - Deduct from Above Prices			20%
Steel Tapered - Deduct from Above Prices			20%
Steel Tube - Deduct from Above Prices			40%
Add for Concrete Bases		360.00	240.00

1012.0 FLOOR & ROOF ACCESS DOORS (M) (Carpenters)

Floor - Single Leaf 1/4" Steel

	UNIT	LABOR	MATERIAL
24" x 24"	Each	101.00	630.00
30" x 30"	Each	127.00	680.00
30" x 36"	Each	156.00	785.00
36" x 36"	Each	180.00	815.00
Deduct for Recess Type	Each		35.00
Add for Aluminum	Each		65.00

Roof - Steel

	UNIT	LABOR	MATERIAL
30" x 36"	Each	160.00	590.00
30" x 54"	Each	170.00	830.00
30" x 96"	Each	250.00	1,510.00
Add for Disappearing Stairs	Each	770.00	5,200.00
Add for Fixed stairs	Each	215.00	1,400.00

1013.0 FLOOR MATS & FRAMES (M) (Carpenters)

	UNIT	LABOR	MATERIAL
Mats - Vinyl Link 4' x 5' x 1/2"	Each	44.00	380.00
4' x 7' x 1/2"	Each	47.00	570.00
Add for Color	SqFt		4.00
Frames - Aluminum 4' x 5'	Each	80.00	235.00
4' x 7'	Each	86.00	290.00

1014.0 FOLDING GATES (M) (Carpenters)

	UNIT	LABOR	MATERIAL
8' x 8' - with Lock Cylinder	Each	310.00	1,450.00
	or SqFt	4.85	22.00

1015.0 LOCKERS (Incl. Benches) (M) (Sheet Metal Workers)

Single - Open Type with 6" Legs

	UNIT	LABOR	MATERIAL 60"	72"
9" x 12" x 72"	Each	22.00	125.00	130.00
9" x 15" x 72"	Each	22.00	130.00	137.00
12" x 12" x 72"	Each	25.00	140.00	148.00
12" x 15" x 72"	Each	26.00	145.00	149.00
12" x 18" x 72"	Each	27.00	147.00	150.00
15" x 15" x 72"	Each	28.00	149.00	153.00
15" x 18" x 72"	Each	29.00	157.00	155.00
15" x 21" x 72"	Each	30.00	162.00	168.00

Double Tier

	UNIT	LABOR	MATERIAL
12" x 12" x 36"	Each	17.25	73.00
12" x 15" x 36"	Each	17.80	76.00
12" x 18" x 36"	Each	18.00	78.00
15" x 15" x 36"	Each	18.00	81.00
15" x 18" x 36"	Each	18.50	88.00
15" x 21" x 36"	Each	19.00	93.00

Box Lockers

	UNIT	LABOR	MATERIAL
12" x 12" x 12"	Each	16.00	46.00
15" x 15" x 12"	Each	16.50	48.00
18" x 18" x 12"	Each	17.25	52.00
12" x 12" x 24"	Each	17.25	62.00
15" x 15" x 24"	Each	17.75	70.00
18" x 18" x 24"	Each	18.25	72.00
Add Flat Key Locks (Master Keyed)	Each	-	10.00
Add for Combination Locks	Each	-	12.00
Add for Sloped Tops	Each	8.10	12.00
Add for Closed Base	Each	8.10	9.00
Add for Other than Standard Colors	Each	-	15%
Add for Benches	LnFt	9.00	30.00
Add for Less than Banks of 4	Each	5%	10%
Add for Recessed Latch (Quiet)	Each	-	3.50

1016.0 LOUVERS AND VENTS (M) (Carpenters or Sheet Metal Workers)

	UNIT	LABOR	MATERIAL
LOUVRES			
Exterior			
Galvanized - 16 Ga	SqFt	15.00	46.00
Aluminum - 14 Ga	SqFt	15.30	55.00
Extruded 12 Ga	SqFt	15.30	63.00
Add for Brass, Bronze or Stainless	SqFt	-	52.00
Add for Insect Screen	SqFt	-	4.50
Interior Door or Partition - Steel	SqFt	14.00	57.00
Aluminum	SqFt	14.00	60.00
VENTS - Brick - Aluminum	Each	28.00	75.00
Block - Aluminum	Each	28.00	115.00
See Div. 15 for Mechanically Connected			

1017.0 PARTITIONS - WIRE MESH (M) (Carpenters)

	UNIT	LABOR	MATERIAL
PARTITIONS			
6 Ga - 2" Mesh & Channel Frame	SqFt	5.30	11.50
10 Ga - 1 1/2" Mesh - 8'	SqFt	4.30	9.10
Add for Door	Each	106.00	425.00
WINDOW GUARDS - 10 Ga with Frame	SqFt	6.25	18.00
Add for Galvanized	-	-	15%

1018.0 PARTITIONS AND CUBICLES - DEMOUNTABLE (L&M) (Carpenters)

	UNIT	COST
PARTITIONS - DEMOUNTABLE - 1 9/16" x 10'		
Aluminum or Enameled Steel Extrusions (Acoustical)		
Gypsum Panels (Unpainted)	SqFt	14.50
Vinyl Laminated Gypsum Panels	SqFt	15.80
Steel Panels	SqFt	15.80
Wood Panels - Hard Wood	SqFt	10.50
Add for 2 3/4"	SqFt	2.00
Add per Door	Each	570.00
CUBICLES - DEMOUNTABLE (Includes Glass)		
Steel Frame and Panels (Acoustical) Burlap		
3' - 6" High	SqFt	24.50
4' - 8" High	SqFt	22.00
5' - 8" High	SqFt	18.80
7' - 0" High	SqFt	19.30
Aluminum Frame and Panels (Acoustical) Burlap		
3' - 6" High	SqFt	19.60
4' - 8" High	SqFt	18.20
5' - 8" High	SqFt	17.80
7' - 0" High	SqFt	17.30
Deduct for Non-Acoustical Type	SqFt	2.15
Add for Vinyl Laminated	SqFt	1.95
Add for Carpeted	SqFt	4.40
Add for Curved Panels	SqFt	7.70
Add per Door	Each	465.00
Add per Gate	Each	290.00

1019.0 PARTITIONS - OPERABLE (Including Moving Dividers and Operable Walls) (L&M) (Carpenters)

	UNIT	LABOR	MATERIAL
ACCORDION FOLD (Fabric)	SqFt	5.40	16.20
Add for Acoustical Type	SqFt	-	5.70
Add for Metal & Wood Supports	LnFt	4.90	7.00
FOLDING (Wood, Plastic Lam., Fabric Cover)	SqFt	6.50	33.50
Add for Acoustical Type	SqFt	-	6.30

		UNIT	LABOR	MATERIAL
1019.0	**PARTITIONS - OPERABLE, Cont'd...**			
	OPERABLE WALLS (Electrically Operated)			
	Folding - Plastic Laminated	SqFt	7.75	43.00
	Vinyl Overlaid	SqFt	7.75	41.00
	Wood (Prefinished)	SqFt	7.75	41.00
	Side Coiling - Wood	SqFt	-	57.00
	Metal	SqFt	-	62.00
	REMOVABLE OR PORTABLE WALLS (Acoustical)			
	4"	SqFt	-	58.00
	2 1/4"	SqFt	-	46.00
1020.0	**PEDESTRIAN CONTROLS - GATES & TURNSTILES**	-	-	-
1021.0	**POSTAL SPECIALTIES (M) (Carpenters)**			
	CHUTES - Aluminum	Per Floor	250.00	1,080.00
	Bronze	Per Floor	250.00	1,540.00
	RECEIVING BOX - Aluminum	Each	145.00	1020.00
	Bronze	Each	145.00	1,700.00
	LETTER BOXES			
	3" x 5" Alum. with Cam Locks Front Load	Each	11.80	69.00
	Rear Load	Each	11.80	84.00
1022.0	**SHOWER & TUB RECEPTORS & DOORS (M) (Cement Finishers)**			
	PRECAST TERRAZZO WITH STAINLESS STEEL CAP			
	32" x 32" x 6"	Each	110.00	350.00
	36" x 36" x 6"	Each	121.00	430.00
	36" x 24" x 12"	Each	125.00	620.00
1023.0	**SIGNS, LETTERS & PLAQUES (M) (Carpenters)**			
	LETTERS			
	Cast Aluminum, Baked Enamel, Anodized Finish			
	6" x 1/2" x 2"	Each	12.80	31.00
	8" x 1/2" x 2"	Each	14.00	37.00
	12" x 1/2" x 2"	Each	15.20	69.00
	Plastic - Character	Each	3.00	4.30
	PLAQUES - 24" x 24" Cast Aluminum	Each	115.00	1,000.00
	24" x 24" Bronze	Each	116.00	1,180.00
	SIGNS - HANDICAP (Metal)	Each	15.00	40.00
1024.0	**STORAGE SHELVING (L&M) (Sheet Metal Workers)**			
	METAL - 18 Ga - 36" Sections 7'1" High			
	12" Deep, 6 Shelves	LnFt	16.00	44.00
	18" Deep, 6 Shelves	LnFt	17.00	52.00
	24" Deep, 6 Shelves	LnFt	19.20	61.00
	Add or Deduct per Shelf	LnFt	3.10	5.50
	Add for Closed Back and Ends	LnFt	5.20	5.50
	Add for 48' Sections	LnFt	5.20	3.90
	Deduct for 22 Ga	LnFt	.60	3.90
	PALLET RACKS - 6' x 36" - 4 Shelves	LnFt	32.50	118.00
1025.0	**SUN CONTROL & PROTECTIVE COVERS (L&M)**	SqFt	5.90	21.00
1026.0	**TELEPHONE AND SOUND BOOTHS (M) (Carpenters)**	Each	245.00	3,800.00
1027.0	**TOILET & BATH ACCESSORIES (L&M) (Plumbers)**			

		UNIT	LABOR	MATERIAL
	BASED ON STAINLESS STEEL			
	Curtain Rods - 60"	Each	21.00	36.00
	Medicine Cabinets - 14" x 24"	Each	35.00	86.00
	18" x 24"	Each	37.00	128.00
	Mirrors - 16" x 10"	Each	22.50	80.00
	18" x 24"	Each	23.50	98.00
	24" x 30"	Each	36.00	118.00
	Add for Shelves	-	-	31.00
	Mop Holders	Each	30.00	82.00
	Paper Dispenser - Surface Mounted	Each	23.50	37.00
	Recessed	Each	30.00	72.00
	Toilet Seat Covers - Recessed	Each	30.00	84.00

		UNIT	LABOR	MATERIAL
1027.0	**TOILET & BATH ACCESSORIES, Cont'd...**			
	Purse Shelf	Each	25.00	72.00
	Robe Hooks	Each	18.30	18.00
	Handicap Grab Bar - 18"	Each	24.50	42.00
	36"	Each	28.00	49.00
	42"	Each	34.50	54.00
	Set of Three	Each	75.00	110.00
	Soap Dispenser - Wall Mount	Each	24.50	53.00
	- with Lather Shelf	Each	25.50	109.00
	Soap and Grab Bar	Each	24.50	60.00
	Toilet Seat Cover Dispenser	Each	27.60	84.00
	Towel Bars - 24"	Each	24.50	36.00
	Towel Dispenser - Recessed	Each	51.50	480.00
	- Surface Mounted	Each	47.00	155.00
	and Disposal - Recessed	Each	47.00	480.00
	Sanitary Napkin Disposal - Recessed	Each	27.00	72.00
	Electric Hand Dryer	Each	70.00	395.00
1028.0	**TOILET COMPARTMENTS (Incl. Shower & Dressing Rooms) (L&M) (Sheet Metal Workers & Plumbers)**			
	PARTITIONS - FLOOR BRACED			
	(Door & 1 Side Panel) Baked Enameled	Each	133.00	460.00
	Stainless Steel	Each	138.00	1,100.00
	Plastic Laminate	Each	143.00	515.00
	Marble	Each	225.00	850.00
	Add for Extra Side Panel - Baked Enameled	Each	80.00	230.00
	Plastic Laminate	Each	80.00	255.00
	Marble	Each	162.00	330.00
	Add for Ceiling Hung	Each	80.00	110.00
	Add for Reinforcing for Handicap Bars (ADA)	Each	-	82.00
	Add for Paper Holders	Each	17.00	39.00
	Urinal Screens - Baked Enameled	Each	97.00	230.00
	Plastic Laminate	Each	98.00	250.00
	Marble	Each	142.00	420.00
	Shower Partitions (No Plumbing)			
	Baked Enameled (with Curtain & Terrazzo Base)			
	32" x 32"	Each	172.00	720.00
	36" x 36"	Each	172.00	660.00
	Fiberglass & Fiberglass Base - 32" x 32"	Each	108.00	300.00
	36" x 36"	Each	108.00	340.00
	Add for Doors	Each	46.00	230.00
	Tub Enclosures			
	Aluminum Frame and Tempered Glass Door	Each	80.00	215.00
	Chrome Frame and Tempered Glass Door	Each	80.00	282.00
1029.0	**URNS AND TRAYS**			
	CIGARETTE URNS - SS Rectangular or Round	Each	45.00	110.00
1030.0	**WALL AND CORNER GUARDS**			
	CORNER GUARDS			
	Stainless Steel			
	16 Ga 4" x 4" with Anchor	LnFt	7.00	23.00
	18 Ga 4" x 4" with Anchor	LnFt	6.40	23.00
	22 Ga 3½" x 3½" with Anchor	LnFt	7.00	19.00
	22 Ga 3½" x 3½" with Tape	LnFt	6.50	18.00
	Aluminum - Anodic with Tape	LnFt	6.50	16.00
	Vinyl	LnFt	6.00	9.80
	See 0502.3 for Steel			

1101.0 MEDICAL CASEWORK (L&M) (Carpenters)
INCLUDING PLASTIC LAMINATE TOP, HARDWARE & FINISH

		UNIT	LABOR	MATERIAL
.1	METAL			
	Base Cabinets - 35" High x 24" Deep	LnFt	35.00	230.00
	Wall Cabinets - 24" High x 12" Deep	LnFt	33.00	180.00
	High Wall (Utility) - 84" High x 12" Deep	LnFt	36.00	185.00
.2	PLASTIC LAMINATE (Including Plastic Laminator)			
	Base Cabinets - 35" High x 24" Deep	LnFt	35.00	230.00
	Wall Cabinets - 24" High x 12" Deep	LnFt	33.00	170.00
	High Wall (Utility) - 84" High x 12" Deep	LnFt	36.00	190.00
.3	WOOD			
	Base Cabinets - 35" High x 24" Deep	LnFt	35.00	215.00
	Wall Cabinets - 24" High x 12" Deep	LnFt	33.00	170.00
	High Wall (Utility) - 84" High x 12" Deep	LnFt	36.00	190.00
.4	STAINLESS STEEL			
	Base - 35" High x 24" Deep	LnFt	36.00	270.00
	Wall - 24" High x 12" Deep	LnFt	33.00	250.00

1102.0 EDUCATIONAL CASEWORK (L&M) (Carpenters)
Same as 1101.0 Medical Casework above.

1103.0 KITCHEN AND BATH CASEWORK (M) (Carpenters)

		UNIT	LABOR	MATERIAL
.1	PREFINISHED WOOD - NO TOPS			
	Base Cabinets - 35" High x 18" Deep	LnFt	35.00	175.00
	Front Only	LnFt	33.00	100.00
	with 1 Drawer	LnFt	35.00	210.00
	with 4 Drawers	LnFt	36.00	280.00
	with Lazy Susan	LnFt	35.50	220.00
	Upper Cabinets - 30" High x 12" Deep	LnFt	34.00	140.00
	18" High x 12" Deep	LnFt	33.00	110.00
	with Lazy Susan	LnFt	34.00	185.00
	High Cabinets - 84" High x 18" Deep	LnFt	36.00	280.00
	Add for Counter Tops - See 0603.2			
.2	PLASTIC LAMINATE			Add 10%

1104.0 MEDICAL EQUIPMENT (M) (Carpenters)

		UNIT	LABOR	MATERIAL
.1	STATION AND SERVICE CENTER EQUIPMENT			
.11	Kitchen -			
	Example: 4'6" x 4'2" Base & 30" Upper	Each	265.00	2,250.00
.12	Medical Preparation -			
	Example: 4'0" x 6'8" x 1'8"	Each	330.00	4,500.00
	(also 5' and 6' wide)			
.13	Nourishment and Ice Stations -			
	Example: 6'0" x 6'8" x 2'8"	Each	350.00	11,800.00
.14	Lavatory Unit -			
	Example: 4'7" x 1'8" x 4'0"	Each	160.00	1,150.00
.15	Janitor's Closet -			
	Example: 2'0" x 5'4" x 6'8"	Each	300.00	4,500.00
.16	Glove Packaging - Example	Each	160.00	1,250.00
.17	Linen Inspection - Example	Each	178.00	1,700.00
.2	LAB TABLES AND DEMONSTRATION DESKS	Each	170.00	2,500.00

1104.0 MEDICAL EQUIPMENT, Cont'd...

		UNIT	LABOR	MATERIAL
.3	**TOPS AND BOWLS (M) (Carpenter or Plumber)**			
	Calcium Aluminum Silicate	SqFt	7.00	26.50
	Corian	SqFt	7.00	51.00
	Epoxy Resins (Kemstone)	SqFt	7.00	24.00
	Marble - Artificial	SqFt	5.40	26.00
	Plastic Laminate - 25" - 1 1/2"	SqFt	4.40	24.00
	Add for Backsplash	LnFt	1.90	5.00
	Soapstone	SqFt	4.90	36.00
	Stainless Steel - Bowl	Each	75.00	205.00
	Tops	SqFt	6.80	68.00
	Add for Backsplash	LnFt	2.30	9.00
	Vinyl Sheet - 25" - 1 1/2"	SqFt	3.40	15.20
	Add for Backsplash	LnFt	1.85	5.00
.4	**FUME HOODS (INCLUDING DUCT WORK)**			
	Stock - 4"	Each	460.00	5,750.00
	Custom	Each	460.00	6,100.00
.5	**KEY CABINETS (100 KEY)**	Each	90.00	330.00
.6	**BLANKET AND SOLUTION WARMERS**			
	(NO MECHANICAL OR ELECTRICAL)			
	Example: 27 x 11 x 30" x 74			
	Free Standing - Steam	Each	420.00	5,300.00
	Electric	Each	420.00	5,500.00
	Recessed - Steam	Each	475.00	5,400.00
	Electric	Each	475.00	5,500.00
.7	**ENVIRONMENTAL GROWTH CHAMBERS & LABS**	Each	1,300.00	5,600.00
.8	**CHART RACKS AND CHART RACK DESKS**			
	Racks	Each	80.00	360.00
	Desks	Each	120.00	1,750.00

			ELECTRIC		STEAM	
		UNIT	Recessed	Open Mtd or Mobile	Recessed	Open Mtd or Mobile
.9	**STERILIZING EQUIPMENT** (NO MECHANICAL/ ELECTRICAL)					
	Dressing and Instrument -					
	Example: 16" x 26"	Each	14,000	-	1,460	-
	Dressing -					
	Example: 20" x 20" x 38"	Each	29,000	-	24,000	2,500
	Solution Warming/ Storage Cab.	Each	-	-	-	-
	Bedpan -					
	Example: 16" x 16" x 27"					
	Non Pressure	Each	-	3,400	-	-
	Pressure	Each	-	3,300	-	-
	Laboratory (Painted)	Each	16,000	18,000	11,000	18,200
	Add for Stainless Steel					
	Autoclave (All Purpose)					
	Portable Inst. - Non Pressure					
	6" x 6" x 13"	Each	-	1,360	-	-
	8" x 8" x 16"	Each	-	2,000	-	-
	10" x 10" x 22"	Each	-	7,000	-	-
	Hot Air -					
	Example: 39" x 14" x 19"					
	Gravity	Each	3,800	3,650	-	-
	Convection	Each	5,200	7,000	-	-

DIVISION #11 - EQUIPMENT

1104.0 MEDICAL EQUIPMENT, Cont'd...

.10 X-RAY AND DARK ROOM EQUIPMENT

Lead Protection - See 0801 for Doors & View Window Frames
Lead Protection - See 0902 for Lead Lined Sheetrock
Manual Dark Room

	UNIT	LABOR	MATERIAL
Example: Room including Cassette Pass Box, Cassette Storage Cabinet, Film Identifier Cabinet, Film Illuminator and Dryer, Development Tank including Tops and Splashes	Each	1,600.00	16,000.00
Automatic			
Processor with Solution Storage Tank	Each	870.00	12,080.00
Film Loader	Each	305.00	6,000.00
.11 AUTOPSY AND MORTUARY EQUIPMENT			
Refrigerators	Each	495.00	10,100.00
Tables	Each	315.00	6,300.00
Carts	Each	89.00	1,690.00
.12 NARCOTIC SAFES	Each	100.00	650.00
.13 REFRIGERATORS (OTHER THAN MORTUARY)			
Under Counter	Each	250.00	3,300.00
.14 TOTE TRAYS AND UTILITY CARTS	Each	86.00	525.00
.15 GAS TRACKS - Example: Operating Room	Each	540.00	3,500.00
Straight	LnFt	48.00	300.00
Curved	LnFt	59.00	370.00
.16 ANIMAL FENCING AND CAGING	SqFt	3.00	6.50
.17 DENTAL	-	-	-
.18 SURGICAL	-	-	-
.19 INCUBATORS	Each	-	4,800.00
.20 THERAPY-HEAT	Each	-	2,350.00
.21 OVERHEAD HOISTS	Each	-	2,300.00

1105.0 EDUCATIONAL EQUIPMENT

1106.0 RESIDENTIAL EQUIPMENT (NO MECH. OR ELEC.)

See 1504 for Supply, Waste and Vent
See 1611 for Electric Power

.1 FREE STANDING

	UNIT	LABOR	MATERIAL
Clothes Dryer - Gas	Each	89.00	500.00
Electric	Each	89.00	470.00
Clothes Washers	Each	89.00	500.00
Compactors - 15"	Each	85.00	550.00
Disposals	Each	85.00	180.00
Dishwasher	Each	89.00	600.00
Freezers - Upright - 16 CuFt	Each	89.00	490.00
19 CuFt	Each	97.00	550.00
21 CuFt	Each	110.00	650.00
Freezers - Upright - Frost Proof			
15 CuFt	Each	89.00	590.00
18 CuFt	Each	89.00	680.00
Freezers - Chest - 15 CuFt	Each	89.00	590.00
20 CuFt	Each	95.00	680.00
25 CuFt	Each	108.00	800.00
Dehumidifier - 40 Pint	Each	90.00	200.00
Humidifier - 10 Gals/Day	Each	90.00	280.00
Range Hood	Each	75.00	195.00
Water Heater - 30 Gal. & 40 Gal.	Each	135.00	260.00
Water Softener - 30 Grains	Each	160.00	850.00

1106.0 RESIDENTIAL EQUIPMENT, Cont'd...

	UNIT	LABOR	MATERIAL
.1 FREE STANDING, Cont'd.			
Refrigerators			
Compact - 11 CuFt	Each	90.00	430.00
Conventional- 14 CuFt	Each	98.00	540.00
With Freezer- 15 CuFt (Also 17 & 19)	Each	100.00	650.00
21 CuFt	Each	105.00	800.00
Frostproof- 12 CuFt	Each	100.00	700.00
15 CuFt	Each	110.00	900.00
Side-by-Side- 22 CuFt	Each	110.00	1,200.00
25 CuFt	Each	112.00	1,450.00
Ranges - Electric - 30"	Each	90.00	700.00
40"	Each	98.00	1,500.00
Gas - 20"	Each	85.00	380.00
30"	Each	90.00	720.00
Microwave Ovens - Counter	Each	90.00	240.00
Over-counter	Each	125.00	400.00
.2 BUILT-IN			
Air Conditioners - 110 Volt	Each	85.00	410.00
220 Volt	Each	85.00	600.00
Dishwashers	Each	85.00	570.00
Disposals	Each	85.00	175.00
Hoods	Each	85.00	190.00
Ranges - Surface - 30" Gas	Each	85.00	650.00
42" Gas with Broiler	Each	88.00	1,400.00
30" Electric	Each	85.00	700.00
42" Elec. with Broiler	Each	88.00	1,400.00
Ovens - Single - Gas	Each	95.00	750.00
Double	Each	105.00	1,000.00
Single - Electric	Each	105.00	700.00
Double	Each	110.00	1100.00

1107.0 ATHLETIC EQUIPMENT (M) (Carpenters)

	UNIT	LABOR	MATERIAL
.1 BASKETBALL BACKSTOPS			
Stationary	Each	390.00	740.00
Retractable	Each	550.00	1,800.00
Suspended	Each	700.00	3,600.00
Add for Electric Operated	Each	-	1,060.00
.2 GYM SEATING - See Div. 1306.0			
.3 SCOREBOARDS (Great Variable)	Each	1100.00	6,400.00
.4 BOWLING AND BILLIARDS - See Div. 1302.0			
.5 SWIMMING POOL EQUIPMENT- See Div. 1318.0			
.6 OUTSIDE RECREATIONAL EQ.- See Div. 0208.2			
.7 GYMNASTIC EQUIPMENT - See Div. 0208.2			
.8 SHOOTING RANGE EQUIPMENT	Point	800.00	6,900.00
.9 HEALTH CLUB EQUIPMENT	-	-	-

1108.0 FOOD SERVICE EQUIPMENT (L&M) (Sheet Metal and Plumbing)

	UNIT		COST
Broilers - 36" Electric	Each	-	3,000.00
Can Washers	Each	-	3,600.00
Coffee Stand and Urn	Each	-	4,000.00
Hot Food Carts	Each	-	3,750.00
Conveyors - 12'	Each	-	6,400.00
Cutter - Mixer	Each	-	12,000.00
Dishtables - Clean	Each	-	4,000.00
Soiled	Each	-	4,800.00

1108.0	FOOD SERVICE EQUIPMENT, Cont'd...	UNIT	COST
	Disposer - Garbage - 3 H.P.	Each	3,000.00
	1 1/2 H.P.	Each	2,400.00
	Dishwasher- Automatic – Conveyor Type	Each	14,000.00
	Door Type	Each	9,000.00
	Fryers - Double	Each	4,920.00
	Hot Food Carts	Each	3,750.00
	Ice Cream Dispenser	Each	1,600.00
	Ice Machine with Bin	Each	4,500.00
	Juice and Beverage Dispenser	Each	4,300.00
	Meat Saw and Grinder	Each	3,700.00
	Microwave	Each	400.00
	Oven - Double Convector - Electric	Each	8,400.00
	Single	Each	4,500.00
	Portable Sink	Each	1050.00
	Range - Oven- Four Burner	Each	1,300.00
	Range Hoods – Per Foot	Per Foot	400.00
	Refrigerators – Reach In	Each	2,900.00
	Silverware Dispensers	Each	300.00
	Sinks - Utility	Each	900.00
	Soiled Dish	Each	3,500.00
	Slicer	Each	3,300.00
	Steamers	Each	11,000.00
	Tray Dispensers	Each	1,000.00
	Setup Unit	Each	1,600.00
	Utility Carts	Each	420.00
	Water Stations	Each	700.00
	Work Tables –30"	Each	800.00
	Fire Protection	SqFt	36.00
1109.0	LAUNDRY EQUIPMENT (L&M)		
	Dryer - 50 Lb.	Each	2,200.00
	Ironer, Air	Each	7,600.00
	Hand	Each	5,200.00
	Washer Tumbler - 135 Lb.	Each	38,000.00
	50 Lb.	Each	21,000.00

1110.0 LIBRARY EQUIPMENT (M) (Carpenters)

			LABOR	MATERIAL
	Book Trucks	Each	100.00	1,300.00
	Card Catalog Index - 30 drawer	Each	130.00	2,250.00
	Chairs	Each	18.50	120.00
	Charging Desks - Card File	Each	100.00	1,140.00
	Charging Unit	Each	100.00	1,220.00
	Open Shelf	Each	100.00	805.00
	Newspaper Rack	Each	100.00	780.00
	Shelving (Maple) - Example: 36" x 60" x 8"	LnFt	19.00	115.00
	Tables	Each	75.00	790.00

1111.0 ECCLESIASTICAL EQUIPMENT (M) (Carpenters)

.1	CROSSES, SPIRES AND STEEPLES	-	-	-
.2	CONFESSIONALS - Two Person	Each	650.00	6,100.00
	One Person	Each	500.00	4,200.00
.3	PEWS (OAK) (With Kneelers)	LnFt	16.00	120.00
	Add for Frontals	LnFt	16.00	60.00
	Add for Padded Kneelers	LnFt	4.30	3.40
	Add for Birch	LnFt	-	12.00
.4	BAPTISMAL FONTS	Each	72.00	850.00
.5	MISCELLANEOUS (WOOD ONLY)			
	Main Altars	Each	485.00	5,980.00
	Side Altars	Each	280.00	2,250.00
	Lecterns	Each	160.00	1,350.00
	Credence Table	Each	135.00	900.00

1112.0 STAGE AND THEATRICAL EQUIPMENT (L&M) (Carpenters)

.1 STAGE DRAPERIES, HARDWARE, CONTROLS
AND OTHER EQUIPMENT

	UNIT	LABOR	MATERIAL
Motorized Rolling Curtain - 38' x 16'	Each	-	6,500.00
Stage Equipment Sets	Each	-	2,300.00
Light Bridge	Each	-	23,000.00
.2 FOLDING STAGES - Portable Thrust	Each	-	15,000.00
.3 REVOLVING STAGES - 32' Dia x 12" High	Each	-	140,000.00
.4 FIXED AUDITORIUM SEATING - Cushioned	Each		205.00
.5 PROJECTION SCREENS			
Motor Oper. - 8' x 8'	Each	150.00	1,650.00
10' x 10'	Each	160.00	1,940.00
Stationary - 20' x 20'	Each	195.00	3,150.00

1113.0 BANK EQUIPMENT (M) (Carpenters)

	UNIT	LABOR	MATERIAL
.1 SAFES	Each	175.00	1,300.00
.2 DRIVE-UP WINDOW UNIT - Including Cash Drawers, Counter, Window, Depository and Lock Box	Each	1,050.00	9,500.00
.3 TELLER UNIT	Each	360.00	2,600.00
Cash Dispensing	Each	3,200.00	43,000.00
.4 NIGHT DEPOSITORY - Envelope	Each	240.00	1,180.00
Incl. Head & Chest	Each	1,050.00	10,090.00
.5 SURVEILLANCE SYSTEMS			
Receiver and Transmitter	Each	-	780.00
Cameras	Each	-	1,140.00
Time Lapse Recorder	Each	-	3,700.00
.6 SAFE DEPOSIT BOX	Each	8.00	80.00
Vault Doors - See Div. 0803.16			
Pneumatic Tube System - See Div. 14			

1114.0 DETENTION EQUIPMENT

1115.0 PARKING EQUIPMENT

	UNIT	LABOR	MATERIAL
.1 AUTOMATIC DEVICES			
Cash Registers	Each	600.00	12,500.00
Clocks - Rate Computing	Each	95.00	950.00
Standards	Each	75.00	725.00
Gate	Each	450.00	3,500.00
Loop Sensor	Each	240.00	1,070.00
Treadle	Each	180.00	440.00
Operator Station - Coin	Each	550.00	5,300.00
Key	Each	560.00	1,150.00
Radio Control - Transmitter	Each	-	2,050.00
Ticket Spitter Station	Each	267.00	6,500.00
.2 BOOTHS - Example: 4' x 6'	Each	330.00	6,300.00
.3 METERS AND POSTS	Each	90.00	350.00

1116.0 LOADING DOCK EQUIPMENT

	UNIT	LABOR	MATERIAL
.1 DOCK BUMPERS			
16" x 12" x 4"	Each	25.00	54.00
24" x 6" x 4"	Each	26.00	55.00
24" x 10" x 4"	Each	27.00	60.00
24" x 12" x 4"	Each	28.00	81.00
36" x 6" x 4"	Each	29.00	72.00
36" x 10" x 4"	Each	30.00	89.00
36" x 12" x 4"	Each	35.00	105.00
Add for Galvanized Angles	Each	-	8.00

		UNIT	LABOR	MATERIAL
1116.0	**LOADING DOCK EQUIPMENT, Cont'd...**			
.2	DOCK BOARDS -			
	5,000# 3' - 6" x 6' - 0"	Each	125.00	1,030.00
	10,000# 4' - 0" x 6' - 0"	Each	140.00	1,325.00
	15,000# 4' - 0" x 6' - 0"	Each	150.00	1,750.00
.3	LEVELERS & ADJUSTABLE RAMPS			
	8' - 0" x 8' (Add Concrete)	Each	245.00	3,500.00
.4	SHELTERS - 10' x 10' with Dock Pad	Each	350.00	2,000.00
	Strip Curtain - 10' x 10' x 8"	Each	100.00	265.00
	Dock Pad - 10' x 10'	Each	195.00	720.00
1117.0	**SERVICE STATION EQUIPMENT**			
.1	REELS - Chassis (No Mech. or Electric)	Each	135.00	710.00
	Air	Each	135.00	590.00
	Water	Each	135.00	610.00
	Motor Oil	Each	135.00	700.00
	A.T.F.	Each	135.00	760.00
	Gear Lube	Each	135.00	690.00
	Combination of Above - 3 Hose	Each	550.00	5,400.00
	Combination of Above - 5 Hose	Each	910.00	6,800.00
.2	PUMPS - Chassis	Each	160.00	620.00
	Gear Oil	Each	160.00	680.00
	A.T.F.	Each	160.00	680.00
	Oil - Motor	Each	120.00	720.00
	Gasoline	Each	360.00	2,200.00
.3	COMPRESSORS - 5 H.P.	Each	360.00	4,600.00
	3 H.P.	Each	350.00	2,250.00
	1 1/2 H.P.	Each	350.00	2,150.00
.4	TIRE CHANGERS - AUTO	Each	450.00	3,160.00
	Exhaust Systems - See Div. 15			
	Hoisting Equipment - See Div. 14			
	Tanks/Air/Water Connections - See Div. 15			
1118.0	**MUSICAL EQUIPMENT**			
1119.0	**CHECKROOM EQUIPMENT**			
	MOTORIZED, 1-Shelf, 16 ft	LnFt	22.00	125.00
	2-Shelf, 16 ft	LnFt	30.00	250.00
1120.0	**AUDIO VISUAL EQUIPMENT**			
	PROJECTION SCREEN	Each	-	135.00
	ELECTRIC CONTROLLED	Each	-	540.00
	TELEVISION CAMERA	Each	-	1,060.00
	MONITOR	Each	-	760.00
	VIDEO TAPE RECORDER	Each	-	1,100.00
1121.0	**INCINERATORS**			
	SMALL - 18" x 18"	Each	250.00	2,200.00
	MEDIUM - 22" x 22"	Each	270.00	3,200.00
	LARGE	Each	405.00	4,480.00
1122.0	**PHOTO AND GRAPHIC ART EQUIPMENT**			
1123.0	**WASTE HANDLING EQUIPMENT**			
	COMPACTORS - Bag - 3 CuYd Hopper	Each	980.00	12,000.00
	Cart	Each	980.00	11,000.00

1201.0 MANUFACTURED CASEWORK (L&M) (Carpenters)
(Including Plastic Laminate Top & Hardware)

		UNIT	LABOR	MATERIAL
.1	MEDICAL CASEWORK			
.11	Metal			
	Base Cabinets - 35" High x 24" Deep	LnFt	36.00	225.00
	Upper Cabinets - 24" High x 12" Deep	LnFt	35.00	180.00
	High Wall - 84" High x 12" Deep	LnFt	36.00	185.00
.12	Plastic Laminate			
	Base Cabinets - 35" High x 24" Deep	LnFt	36.00	240.00
	Upper Cabinets - 24" High x 12" Deep	LnFt	35.00	175.00
	High Wall - 84" High x 12" Deep	LnFt	36.00	280.00
.13	Wood - Prefinished			
	Base Cabinets - 35" High x 24" Deep	LnFt	36.00	225.00
	Upper Cabinets - 24" High x 12" Deep	LnFt	35.00	170.00
	High Wall - 84" High x 12" Deep	LnFt	36.00	275.00
.14	Stainless Steel			
	Base - 35" High x 24" Deep	LnFt	37.00	280.00
	Wall - 24" High x 12" Deep	LnFt	36.00	255.00
.2	EDUCATIONAL CASEWORK			

		UNIT	COST
1202.0	**BLINDS, SHADES AND SHUTTERS (L&M) (Carpenters)**		
.1	BLINDS - Horizontal		
	Wood 1"	SqFt	6.60
	Aluminum 1"	SqFt	6.20
	1"	SqFt	4.90
	Steel 2"	SqFt	4.50
	Cloth 1"	SqFt	7.30
	Vertical		
	Cloth or PVC 3"	SqFt	6.00
	Aluminum 3"	SqFt	7.55
.2	SHADES - Cotton	SqFt	2.15
	Vinyl Coated	SqFt	2.47
	Lightproof	SqFt	3.15
	Slat	SqFt	2.52
	Woven Aluminum	SqFt	4.45
	Fibre Glass	SqFt	3.10
	X-ray and Dark Room	SqFt	10.90
.3	SHUTTERS		
	16" x 1 1/8" x 48"	Pair	118.00
	16" x 1 1/8" x 60"	Pair	136.00
	16" x 1 1/8" x 72"	Pair	142.00
1203.0	**DRAPERIES AND CURTAIN HARDWARE (L&M) (Carpenters)**		
	TRACK	LnFt	5.30
	DRAPERIES	SqYd	9.50
	Add for Lined	SqYd	2.10
	Add for Rods	SqYd	2.40
	Add for Motorized	SqYd	7.20
1204.0	**OPEN OFFICE FURNITURE**		

		UNIT	LABOR	MATERIAL
1205.0	**FLOOR MATS AND FRAMES (M) (Carpenters)**			
	MATS - Vinyl Link - 4' x 5' x 1/2"	Each	35.40	355.00
	4' x 7' x 1/2"	Each	35.40	500.00
	Add for Color	SqFt	-	17.60
	FRAMES - Aluminum - 4' x 5'	Each	72.00	220.00
	4' x 7'	Each	83.00	280.00

		UNIT	COST
1206.0	**AUDITORIUM AND THEATRE SEATING (L&M) (Carpenters)**		
	WOOD AND PLYWOOD	Each	165.00
	CUSHIONED	Each	195.00
	ROCKING	Each	275.00
1207.0	**MULTIPLE-USE SEATING**		
	MOVABLE - Chrome and Solid Plastic	Each	79.00
	Wood and Plywood	Each	70.00
1208.0	**BUILT-IN FOLDING TABLES AND SEATING (M) (Carpenters)**		
.1	TABLES - 3' x 8' - Tempered Hardboard Top	Each	215.00
	Vinyl Laminate Top	Each	240.00
	Folding Table & Bench-Pocket Unit	Each	1,800.00
.2	CHAIRS - Metal	Each	29.00

1301.0 AIR SUPPORTED STRUCTURES

		UNIT	COST
.1	TENNIS COURTS - 2 LAYER THERMAL DOMES		
	(Includes Doors, Heating, Equipment, Lights & Power)		
	1 Court - 58' x 118'	SqFt	20.00
	2 Courts - 106' x 118'	SqFt	19.00
	3 Courts - 156' x 118'	SqFt	18.00
	4 Courts - 201' x 118'	SqFt	17.00
	Deduct for Non Thermal Type	SqFt	2.30
.2	CONSTRUCTION DOMES, WAREHOUSES, ETC.	SqFt	8.90
	Add for Heating	SqFt	3.80
	Add for Light and Power	SqFt	2.00
	Add for Air Lock Doors	Each	15,000.00
	Add for Revolving Doors	Each	7,800.00

1302.0 BOWLING ALLEYS (L&M) (Carpenters)

	UNIT	COST
BOWLING ALLEYS (L&M) (Carpenters)	Lane	40,000.00
Add for Automatic Scorers	Lane	16,000.00

1303.0 CHIMNEYS - SPECIAL (Foundations Not Included)

.1 JOB CONSTRUCTED (L&M) (Bricklayers)

Inside Diam.	Height			COST
5'	100'	Radial Brick	Each	135,000.00
6'	150'	Radial Brick	Each	165,000.00
7'	200'	Radial Brick	Each	217,000.00
8'	250'	Reinforced Concrete	Each	550,000.00
9'	200'	Reinforced Concrete	Each	670,000.00
10'	250'	Reinforced Concrete	Each	900,000.00
12'	300'	Reinforced Concrete	Each	1,600,000.00

.2 PREFABRICATED (METAL REFRACTORY LINED)
1800° - 2000° (L&M) (Sheet Metal & Iron Workers)

	UNIT	COST
10"	LnFt	66.00
12"	LnFt	70.00
15"	LnFt	83.00
18"	LnFt	100.00
21"	LnFt	121.00
24"	LnFt	145.00
30"	LnFt	260.00
36"	LnFt	370.00
Add for Cleanouts or T's	Each	380.00
Add for V.L. Label	Each	10%
Deduct for 800° Chimney	Each	5%

1304.0 CLEAN ROOMS

	UNIT	COST
CLEAN ROOMS	SqFt	210.00 – 400.00

1305.0 FLOORS - ACCESS OR PEDESTAL (L&M) (Carpenters)
(Includes Laminate Topping)

	UNIT	COST
Aluminum	SqFt	21.50
Steel - Galvanized	SqFt	14.40
Plywood - Metal Covered	SqFt	13.30
Add for Carpeting	SqFt	3.45
Add for Stairs and Ramps	SqFt	50%

1306.0 GRANDSTANDS, BLEACHERS AND GYM SEATING

		UNIT	COST
.1	GRANDSTANDS AND BLEACHERS		
	Not Permanent	Each Seat	28.00
	Permanent - Aluminum	Each Seat	108.00
	Steel	Each Seat	70.00
	Concrete	Each Seat	260.00
.2	GYM SEATING - FOLDING	Each Seat	78.00

1307.0 GREENHOUSE (L&M) (Glaziers)

	UNIT	COST
SINGLE ROOM TYPE - Commercial	SqFt	24.00
Educational	SqFt	32.50
Residential	SqFt	36.00
Add for Heating, Lighting and Benches	SqFt	20.00
Add for Cooling	SqFt	8.50

1308.0 INCINERATORS (L&M) (Ironworkers and Bricklayers)

Stack and Breaching Not Included
Pounds/Hour

	UNIT	COST
100	Each	9,200.00
200	Each	9,300.00
300	Each	10,600.00
400	Each	12,000.00
500	Each	13,200.00
600	Each	16,600.00
800	Each	27,300.00
Add for Gas Burner	Each	1,850.00
Add for Controls	Each	980.00
Add for Piping	Each	1,070.00

1309.0 INSULATED ROOMS (L&M) (Carpenters)

		UNIT	COST
.1	CUSTOM - Cooler	SqFt	190.00
	Freezer	SqFt	220.00
.2	PREFABRICATED (SELF-CONTAINED)		

	UNIT	COST
Cooler Example: Room 7'6" x 9'6"		
Wood (Plywood)	Each	7,200.00
Metal Covered 20 ga Galvanized #3 SS	Each	8,900.00
Add for 3/4 HP Compressor	Each	1,180.00
Add for 3/4 HP Compressor Starter & Coils	Each	5,250.00
Add for Electrical	Each	950.00
Cooler - Total Cost - Wood	SqFt	200.00
Metal Covered	SqFt	220.00
Freezer Example: Room 7'6" x 9'6"		
Wood (Plywood)	Each	7,500.00
Metal Covered 20 ga Galvanized #3 SS	Each	9,200.00
Add for 3/4 HP Compressor	Each	1,550.00
Add for 3/4 HP Compressor Starter & Coils	Each	4,750.00
Add for Electrical	Each	980.00
Freezer - Total Cost - Wood	SqFt	220.00
Metal Covered	SqFt	250.00
Add for Shelving	SqFt	10.00

		UNIT	COST
1310.0	**METAL BUILDINGS**		
.1	SHELL - FLOOR AREA (NO FOUNDATION)	SqFt	16.20
	24 ga. 12' Sidewalls and 60' Span (Galvanized)		
	Add per foot above 12' - Floor Area	SqFt	.45
	Add per foot of Span above or below 60'	SqFt	.28
	Add for 2' Overhand - Overhang Area	SqFt	6.30
	Add for Aluminum	SqFt	.70
	Add for Liner Panels - Wall Area	SqFt	2.25
	Add for Insulation 3 1/2" - Wall Area	SqFt	.88
	Add for Insulation 6" - Ceiling Area	SqFt	1.10
	Add for Door Openings - Including Hardware	Each	790.00
	Add for Window Openings	Each	410.00
	Add for Insulated Glass Area	SqFt	11.50
	Add for Overhead Door Area	SqFt	11.00
	See 0803.4 for Door Cost		
	Add for Gutter	LnFt	5.40
	FOUNDATION & GROUND SLAB COST AVG (3-Foot Frost)	SqFt	9.00
1311.0	**RADIATION PROTECTION (L&M) (Carpenters)**		
.1	ELECTROSTATIC SHIELDING		
	Custom - Vinyl Sandwiched Copper Mesh for Floors, Walls & Ceilings (including Copper Foil Tabs & Plastic Tape)	SqFt	27.50
	Add per door and Frame with RF Shielded Glass (including Grounding Wires)	Each	4,100.00
	Add per Borrowed Light RF Shielded Glass	Each	1,700.00
	Example: Room 10' x 8' (Prefab)	Room	33,000.00
		or SqFt	400.00
.2	ELECTROMAGNETIC SHIELDING		
	Example: Room 10' x 8' (Prefab)	Room	13,800.00
		or SqFt	174.00
	Add for Finish, Electricity and Erection	Room	11,500.00
.3	X-RAY		
	Deep Therapy		
	Example: Room 10' x 15'	Room	13,900.00
	Radiography or Fluoroscopy		
	Example: Room 10' x 15'	Room	6,000.00
.4	RADIO FREQUENCY SHIELDING		
	Example: Room 10' x 10'	Room	34,500.00
	See 1104.10 for X-Ray Equipment		
1312.0	**SAUNAS (M) (Carpenters)**		
	Examples (includes Heaters):		
	Room 5' x 6' Cedar (5 K.W. Unit)	Room	5,350.00
		or SqFt	175.00
	Room 6' x 6' Cedar (6 K.W. Unit)	Room	6,400.00
		or SqFt	180.00
	Room 6' x 7' Cedar (7.5 K.W. Unit)	Room	6,480.00
		or SqFt	156.50
	Add for Electrical Connections and Ventilating		925.00

1313.0 SCALES (NO PORTABLE) (Ironworkers)
.1 FLOOR SCALES (HEAVY DUTY INDUSTRIAL)

Size	Capacity	Labor	Material	Concrete Work
46" x 38"	800 lb.	480.00	4,950.00	-
48" x 48"	1,350 lb.	510.00	5,250.00	-
60" x 48"	1,600 lb.	725.00	5,650.00	-
72" x 48"	2,600 lb.	750.00	6,600.00	-
76" x 54"	6,000 lb.	1,300.00	6,900.00	-
84" x 60"	10,000 lb.	2,150.00	9,100.00	-

.2 TRUCK SCALES - FULL WITH WEIGH BEAM
Truck (4-Section) -

Size	Capacity	Labor	Material	Concrete Work
45" x 10"	50 Ton	3,100.00	24,600.00	23,100.00
50" x 10"	50 Ton	3,300.00	25,200.00	23,700.00
60" x 10"	50 Ton	3,800.00	26,400.00	24,200.00
70" x 10"	50 Ton	4,540.00	27,400.00	26,000.00
Add for 60-Ton Capacity			3,900.00	

Axle Load -

Size	Capacity	Labor	Material	Concrete Work
8" x 10"	20 Ton	2,150.00	11,300.00	7,750.00
8" x 10"	30 Ton	2,800.00	12,400.00	7,850.00
10" x 10"	20 Ton	3,000.00	12,900.00	8,300.00
10" x 10"	30 Ton	3,500.00	14,500.00	8,400.00
Add to Above for Printer and Dial			3,000.00	

.3 CRANE SCALES -

	Capacity	Labor	Material	
	1 Ton	165.00	2,950.00	-
	5 Ton	320.00	3,200.00	-

1314.0 SIGNS (Special Exterior) (L&M) (Sheet Metal Workers)

1315.0 SKYDOMES (L&M) (Carpenters)
EXTRUDED ALUMINUM FRAME, WIRE GLASS
EXTERIOR AND FIBERGLASS SUB CEILING

	UNIT	COST
12' x 18' Ridge Type	Each	9,100.00
	or SqFt	42.00
18' x 18' Pyramid Type	Each	17,500.00
	or SqFt	54.00
36' Diameter Geometric Domes	Each	72,000.00
	or SqFt	69.00
18' Diameter Revolving Domes	Each	18,540.00
	or SqFt	66.00

1316.0 SOUND INSULATED ROOMS (Including Vibration Control) (M) (Carpenters)
.1 ANECHOIC ROOMS (ISOLATORS, PANELS AND WEDGES)

	UNIT	COST
ANECHOIC ROOMS	Room	135,000.00
Example: Room 30' x 30'	or SqFt	150.00

.2 AUDIOMETRIC ROOM (PREFABRICATED PANELS)

	UNIT	COST
Example: Research Room - 5' x 8'	Room	12,000.00
	or SqFt	300.00
Example: Medical Practice Room - 5' x 8'	Room	27,000.00
	or SqFt	675.00

1318.0 SWIMMING POOLS

.1 POOLS

Deep Pools - 3 to 9 feet
(Incl. Filters, Skimmers, Lights, Clng. Equip.)

	UNIT	COST
Rectangular or L Shape:		
Residential - 1,000 SqFt	Each	43,500.00
	or SqFt	43.00
Motels and Apartments - 2,000 SqFt	Each	100,000.00
	or SqFt	50.00
Municipal - 5,000 SqFt	Each	285,000.00
	or SqFt	57.00
Add for Diving Stand - Steel - 1 Meter	Each	3,600.00
Steel - 3 Meter	Each	5,350.00
Add for Diving Boards - 12' Fiberglass	Each	930.00
16' Aluminum	Each	1,700.00
Add for Slides - 6' Fiberglass	Each	900.00
12' Fiberglass	Each	1,400.00
Add for Ladders - 10'	Each	350.00
Add for Lifeguard Chairs	Each	1,160.00
Add for Covers	Each	830.00
Shallow Pools - 3 to 5 feet - Gunite with Plaster Finish		
Residential - 1,000 SqFt	Each	32,000.00
	or SqFt	32.00
Motels and Apartments - 2,000 SqFt	Each	84,000.00
	or SqFt	42.40
Deduct for Vinyl Lined Pools	SqFt	10.00
Add for Free Form Pools	Each	5%
Add for Heaters	Each	4,300.00
Add for Automatic Cleaners	Each	4,100.00
Wading Pools - 12' x 12' or 9' Round	Each	5,200.00
	or SqFt	36.40

.2 DECKS - CONCRETE - Average 10' around Pool

	UNIT	COST
and Average 1,000 SqFt	Each	4,000.00
	or SqFt	4.00
Add for Tile	SqFt	9.00
Add for Brick	SqFt	7.60

.3 FENCES - 4' ALUMINUM OR GALVANIZED

	UNIT	COST
	LnFt	1,220.00

1319.0 TANKS

.1 STEEL - GROUND LEVEL (NO FOUNDATION)

	UNIT	COST
150,000 Gals	Each	150,500.00
500,000 Gals	Each	255,000.00
1,000,000 Gals	Each	420,000.00
2,000,000 Gals	Each	670,000.00

.2 STEEL - ELEVATED

	UNIT	COST
100,000 Gals	Each	335,000.00
250,000 Gals	Each	445,500.00
500,000 Gals	Each	725,000.00
1,000,000 Gals	Each	1,650,000.00

1320.0 WOOD DECKS - WITH RAILS

	UNIT	COST
CEDAR	SqFt	16.60
TREATED FIR	SqFt	14.80
Add for Seats	LnFt	4.10
Add for Stairs	Riser	57.00

1401.0 ELEVATORS (L&M) (Elevator Constructors & Ironworkers)

.1 ELECTRIC (CABLE/ TRACTION)

.11 Passenger - Std Specifications

	Speed to F.P.M.	2,000#	3,000#	4,000#	10,000#
Geared A.C. Rheostatic: Selective Collective Operation					
3 Openings	125	74,000	-	-	-
4 Openings	125	77,000	-	-	-
5 Openings	125	80,000	-	-	-
Add per Floor	-	5,500	-	-	-
Geared Variable Voltage: Selective Collective Operation					
8 Openings	250	133,000	140,000	143,000	-
	350	137,000	141,000	146,000	-
10 Openings	250	136,000	140,000	148,000	-
	350	149,000	138,000	151,000	-
Add per Floor	-	7,100	7,600	8,100	-
Group Supervisory					
8 Openings	250	-	150,000	152,000	-
	350	-	158,000	157,000	-
10 Openings	250	-	150,000	155,000	-
	350	-	157,000	158,000	-
Add per Floor	-	-	7,500	8,000	-
Gearless Variable Voltage: Group Supervisory					
8 Openings	500	-	184,000	179,000	-
	800	-	196,000	193,000	-
	1000	-	207,000	210,000	-
10 Openings	500	-	184,000	189,000	-
	800	-	203,000	208,000	-
	1000	-	221,000	211,000	-
Add per Floor	-	-	7,400	7,600	-
Add per Fl. of Express Zone	-	6,900	7,600	-	-

.12 Freight

	Speed to F.P.M.	2,000#	3,000#	4,000#	10,000#
3 Openings	100	-	-	98,000	103,000
4 Openings	100	-	-	103,000	108,000
Add per Floor	-	-	-	23,000	13,200
Add per Power Operated Door	-	-	-	-	7,700

.2 OIL HYDRAULIC

.21 Passenger

	Speed to F.P.M.	2,000#	3,000#	4,000#	10,000#
2 Openings	100	53,000	-	-	-
3 Openings	100	57,000	-	-	-
4 Openings	100	64,000	-	-	-
5 Openings	100	72,000	-	-	-
6 Openings	100	82,000	-	-	-

Deduct for Limited Use Elevator, non-Commercial 8,000# - $5,000

.22 Freight

	Speed to F.P.M.	2,000#	3,000#	4,000#	10,000#
2 Openings	100	-	-	62,000	77,000
3 Openings	100	-	-	73,000	82,000
4 Openings	100	-	-	75,000	85,000

Add per Power Operated Door - $11,000

.3 RESIDENTIAL (WINDING DRUM)

2 Openings - Speed to 36 F.P.M., 700# Capacity - $22,000
   Speed to 30 F.P.M., 410# Capacity - $17,500
Add for 3 Openings - $4,000

Add for ADA Updating -$14,000
Add for Updating Generators, Controllers & Operating Equipment -$60,000
Add to 1401.1 and 1401.2 Above for Custom Specifications - 10%
Add to 1401.2 for Shaft Work, Electric Power & Mechanical - approx. $60,000
Add for Demolition – Electric (5 openings) $18,000
Add for Demolition – Hydraulic (5 openings) $11,000

1402.0 MOVING STAIRS & WALKS (L&M) (Elevator Constructors & Ironworkers)

.1 MOVING STAIRS (Escalators)

16' Floor Height	UNIT	CAPACITY	COST
Metal Balustrade 32" Wide			128,000
48"			138,000
Glass Balustrade 32" Wide			137,000
48"			150,000
Add or Deduct per Foot Floor Height			3,300
Add for Over 20 Feet Floor Height			25%

.2 MOVING WALKS - Belt Type 32"

	UNIT	CAPACITY	COST
MOVING WALKS - Belt Type 32"	LnFt		820
48"	LnFt		930
Pallet Type 32"	LnFt		1,020
48"	LnFt		1,050

1403.0 DUMBWAITERS (L&M) (Elevator Constructors)

.1 ELECTRIC

Drum Type	Speed F.P.M.		COST
2 Openings - 1 Stop	25 to 75	50# to 500#	16,000
3 Openings - 2 Stops	25 to 75	50# to 500#	19,400
Add per Floor			4,200
Traction Type			
3 Openings - 2 Stops	50 to 300	50# to 500#	21,000
4 Openings - 3 Stops	50 to 300	50# to 500#	24,000
5 Openings - 4 Stops	50 to 300	50# to 500#	29,000
6 Openings - 5 Stops	50 to 300	50# to 500#	32,000
Add per Floor			4,400

.2 HAND OPERATED

	COST
2 Openings - 1 Stop	7,000
3 Openings - 2 Stops	7,800
Add per Floor	1,050
Add for all Stainless Steel	25%

1404.0 LIFTS (L&M) (Ironworkers & Elevator Constructors)

		UNIT	CAPACITY	COST
.1	WHEELCHAIR LIFTS - 42"	Each	1,050#	19,200
	96"	Each	1,400#	23,300
.2	STAIR LIFTS (CLIMBERS)	Each		6,600
.3	VEHICLE - SEMI HYDRAULIC LIFTS - 1 Post	Each	8,000#	8,000
	2 Post	Each	10,000#	12,800
	2 Post	Each	15,000#	19,500
	Add for Fully Hydraulic - 2 Post	Each	10,000#	4,450
.4	MATERIAL HANDLING LIFTS	Each		-
.5	STAGE LIFTS - 12' x 30'	Each		94,000
	Add for Permits, Inspections & Elec Hookups			1,600

See 1116 for Loading Dock Lift Equipment

1405.0 PNEUMATIC TUBE SYSTEMS (L&M) (Sheet Metal Workers)

	UNIT	COST
3" ROUND - Single Tube	LnFt	33
Twin Tube	LnFt	43
4" ROUND - Single Tube	LnFt	34
Twin Tube	LnFt	45
4" x 7" OVAL - Single Tube	LnFt	40
Twin Tube	LnFt	57

1406.0 MATERIAL HANDLING (L&M) (Ironworkers & Millwrights)   -

1407.0 TURNTABLES (L&M) (Ironworkers & Millwrights)   -

1408.0 POWERED SCAFFOLDING (L&M) (Ironworkers)   -

Costs of mechanical work are priced as total contractor's costs with an overhead and fee of 20% included. Not included are scaffold, hoisting, temporary heat, enclosures and bonding. Included are 35% taxes and insurance on labor, equipment, tools and 5% sales tax.

1501.0 BASE MATERIALS AND METHODS

		UNIT	COST
.1	PIPE AND PIPE FITTINGS (Including Fittings)		
	Steel - Black - Threaded 3/4"	LnFt	3.50
	1"	LnFt	4.00
	1-1/2"	LnFt	5.80
	2"	LnFt	11.00
	2-1/2"	LnFt	15.50
	3"	LnFt	21.30
	4"	LnFt	26.70
	5"	LnFt	90.50
	6"	LnFt	40.00
	8"	LnFt	57.40
	Steel - Galvanized Schedule 40 3/4"	LnFt	7.30
	1"	LnFt	9.40
	1-1/4"	LnFt	10.60
	1-1/2"	LnFt	12.40
	2"	LnFt	16.20
	3"	LnFt	27.00
	4"	LnFt	35.50
	5"	LnFt	44.00
	6"	LnFt	59.30
	8"	LnFt	65.00
	Deduct for Black Iron	LnFt	5%
	Deduct for Underground	LnFt	10%
	Cast Iron - Hub and Heavy Duty 2"	LnFt	17.00
	3"	LnFt	17.50
	4"	LnFt	24.30
	5"	LnFt	29.05
	6"	LnFt	40.00
	8"	LnFt	47.20
	Add for Line Pipe	LnFt	4.80
	Deduct for Hubless Type	LnFt	10%
	Deduct for Standard	LnFt	15%
	P.V.C. - Schedule 40 1/2"	LnFt	5.10
	3/4"	LnFt	5.60
	1"	LnFt	6.25
	1-1/4"	LnFt	6.80
	1-1/2"	LnFt	7.50
	2"	LnFt	8.20
	3"	LnFt	9.30
	4"	LnFt	15.90
	6"	LnFt	27.30
	Deduct for Underground	LnFt	10%
	P.V.C. - Type L 3/8"	LnFt	3.80
	1/2"	LnFt	5.00
	3/4"	LnFt	5.80
	1"	LnFt	6.80
	1-1/4"	LnFt	7.55
	1-1/2"	LnFt	8.40
	2"	LnFt	9.40
	3"	LnFt	11.00
	4"	LnFt	16.30
	6"	LnFt	31.00
	Add for Type K (Underground)	LnFt	20%
	Deduct for Type M (Residential)	LnFt	5%

1501.0 BASE MATERIALS AND METHODS (Cont'd...)

.1 PIPE AND PIPE FITTINGS (Cont'd...)

	UNIT	COST
Brass - Threadless 1/2"	LnFt	9.70
3/4"	LnFt	11.00
1"	LnFt	13.55
1-1/4"	LnFt	16.70
1-1/2"	LnFt	19.80
2"	LnFt	24.50
3"	LnFt	34.40
4"	LnFt	45.00
Add for Threaded	LnFt	30%
Stainless Steel Schedule 40 1/4"	LnFt	8.90
1/2"	LnFt	13.30
3/4"	LnFt	17.00
1"	LnFt	20.00
2"	LnFt	27.25
4"	LnFt	67.00
Add for Type 316	LnFt	20%
Fibre Glass - 1"	LnFt	17.70
1-1/2"	LnFt	20.30
2"	LnFt	23.00
3"	LnFt	26.70
4"	LnFt	30.50
5"	LnFt	38.50
6"	LnFt	57.60
8"	LnFt	68.00

Utility Piping -

	UNIT	Reinforced Concrete	Vitrified Clay	Bituminous Coated Corrugated	Smooth Paved Corrugated	Ductile Iron
6"	LnFt	$7.30	$7.00	$9.65	$10.00	$16.00
8"	LnFt	7.80	9.50	10.00	11.00	21.60
1"	LnFt	10.00	12.50	11.80	13.10	25.20
12"	LnFt	13.60	14.80	15.00	16.00	39.00
15"	LnFt	15.75	22.30	17.30	18.40	47.50
18"	LnFt	21.30	34.00	21.00	32.00	61.80
21"	LnFt	26.10	40.00	25.10	27.80	75.00
24"	LnFt	29.80	57.00	27.60	32.60	91.00
27"	LnFt	40.25	79.00	33.60	38.00	-
30"	LnFt	49.50	104.00	34.90	44.00	-
36"	LnFt	59.60	144.00	38.20	55.50	-
42"	LnFt	77.00	-	45.20	64.00	-
48"	LnFt	100.00	-	62.10	72.00	-
54"	LnFt	115.00	-	81.20	97.00	-
60"	LnFt	136.00	-	99.00	119.00	-

Add for Trenching and Backfill - See Division 0202.0
See 0205 for Site Drainage Items

1501.0 BASE MATERIALS AND METHODS, Cont'd...

.2 PIPING SPECIALTIES

	UNIT	COST
.21 Expansion Joints - Copper 1/2"	Each	$68.00
3/4"	Each	113.00
1"	Each	159.00
1-1/2"	Each	340.00
2"	Each	545.00
3"	Each	970.00
4"	Each	1,050.00
.22 Vacuum Breakers - 1/2"	Each	41.00
3/4"	Each	47.00
1"	Each	64.00
1-1/2"	Each	86.00

	UNIT	IRON BODY	BRONZE BODY
.23 Strainers - Screwed Ends - 1/2"	Each	34.50	44.80
3/4"	Each	40.00	51.50
1"	Each	46.50	69.00
1-1/2"	Each	64.00	108.00
2"	Each	103.00	180.00
3"	Each	183.00	345.00
4"	Each	485.00	1,320.00
Strainers - Flanged Ends - 2"	Each	164.00	460.00
3"	Each	290.00	950.00
4"	Each	475.00	2,000.00
5"	Each	720.00	2,700.00
6"	Each	990.00	3,150.00

.3 MECHANICAL - SUPPORTING ANCHORS & SEALS

.4 VALVES

Bronze	UNIT	150 psi Gate	150 psi Check	150 psi Globe	Relief	300 psi to 75 Reducing
Body - 1/2"	Each	$31.00	$28.00	$37.00	$56.00	$75.00
3/4"	Each	38.00	31.00	48.00	64.00	90.00
1"	Each	45.00	37.00	66.00	95.00	130.00
1-1/2"	Each	73.00	60.00	102.00	330.00	295.00
2"	Each	95.00	90.00	148.00	360.00	437.00
2-1/2"	Each	170.00	535.00	255.00	-	-
3"	Each	230.00	198.00	380.00	-	-

Iron	UNIT	125 psi Gate	125 psi Check	125 psi Globe
Body - 2"	Each	$272.00	$215.00	$390.00
2-1/2"	Each	283.00	220.00	450.00
3"	Each	318.00	225.00	525.00
4"	Each	482.00	410.00	700.00
6"	Each	670.00	435.00	890.00

		Gate	Check	Globe
Steel - 2"	Each	$910.00	$890.00	$1,150.00
2-1/2"	Each	1,430.00	1,180.00	1,670.00
3"	Each	1,450.00	1,190.00	1,780.00
4"	Each	1,710.00	1,660.00	2,500.00

		Ball	Foot
Plastic - 1/2"	Each	$25.00	$58.00
3/4"	Each	30.00	65.00
1"	Each	35.00	85.00
1-1/2"	Each	60.00	136.00
2"	Each	68.00	163.00

		UNIT	COST
1501.0	**BASE MATERIALS AND METHODS, Cont'd...**		
.5	PUMPS		
	Condensate Return - Simplex	Each	$ 1,900.00
	Duplex	Each	5,100.00
	Utility Water Pump -		
	Multi - Two Stage 3" & 4" 75 HP	Each	19,000.00
	Four Stage 3" & 4" 150 HP	Each	35,500.00
	Single - End Suction 1" x 2" 3 HP	Each	5,000.00
	End Suction 2" x 3" 15 HP	Each	6,320.00
	Condenser Water Pump, Pressure - 100 GPM	Each	16,900.00
	300 GPM	Each	22,600.00
	1200 GPM	Each	51,500.00
	2000 GPM	Each	59,600.00
	Submersible Pump 2"	Each	690.00
	Pedestal Pump	Each	215.00
.6	VIBRATION ISOLATION 5" long	Each	355.00
	6" long	Each	690.00
.7	METERS AND GAGES		
	Disc Type - 3/4"	Each	180.00
	1"	Each	220.00
	2"	Each	620.00
	Compound Type - 3"	Each	2,160.00
	4"	Each	3,360.00
.8	TANKS		
	Fuel Storage - Steel - 1,000 Gallons	Each	2,110.00
	2,000	Each	2,950.00
	5,000	Each	6,700.00
	10,000	Each	9,600.00
	Fuel Storage - Fiberglass - 2,000 Gallons	Each	3,400.00
	5,000	Each	5,900.00
	10,000	Each	9,900.00
	Hot Water Storage - 80 Gallons	Each	405.00
	140	Each	620.00
	200	Each	850.00
	400	Each	1,365.00
	600	Each	2,060.00
	Liquid Expansion - 30 Gallons	Each	360.00
	60	Each	580.00
	100	Each	795.00

1502.0 MECHANICAL INSULATION

.1 INSULATION FOR PIPING

	Unit	3/8" Foam	1" Wall Fiberglass	2" Wall Fiberglass
Interior - Pipe 1/2"	LnFt	3.70	4.55	6.40
3/4"	LnFt	3.80	4.80	6.85
1"	LnFt	4.15	4.05	6.95
1-1/4"	LnFt	4.40	5.45	7.20
1-1/2"	LnFt	4.65	5.90	7.60
2"	LnFt	5.10	6.20	8.50
2-1/2"	LnFt	5.55	6.60	9.20
3"	LnFt	5.95	7.10	10.70
6"	LnFt	8.50	9.10	14.70
8"	LnFt	10.40	13.60	16.30

Exterior - with Aluminum Jacket - Add 2.00/ LnFt

.2 INSULATION FOR DUCT WORK

	Unit	1"	1-1/2"	2"
Rigid with Vapor Barrier	SqFt	2.40	3.35	3.75
Blanket	SqFt	4.95	3.70	3.00

.3 INSULATION FOR BOILERS

	Unit	1"	1-1/2"	2"
Fiberglass - 2"	SqFt	-	-	11.40
Calcium Silicate - 2" with C.F.	SqFt	-	-	17.30

		UNIT	COST
1503.0	**FIRE PROTECTION EQUIPMENT**		
.1	AUTOMATIC SPRINKLER EQUIPMENT - Wet System	SqFt	2.25
		(or) Head	215.00
	Dry System	SqFt	2.60
		(or) Head	230.00
	Add for Pendant Type Sprinklers	SqFt	.65
	Add for Deluge System	SqFt	2.15
.2	CARBON DIOXIDE EQUIPMENT - Cylinder	Each	1,030.00
	Detector & Nozzle	Each	345.00
.3	STANDPIPE AND FIRE HOSE EQUIPMENT		
	Standpipe - Exposed 6"	Each	645.00
	Concealed 6"	Each	825.00
	Cabinets - 30" x 30" Aluminum	Each	330.00
	Hose and Nozzle - 1 1/2"	LnFt	3.00
	2 1/2"	LnFt	3.60
1504.0	**PLUMBING**		
.1	EQUIPMENT	-	-
.2	PACKAGE WASTE, VENT OR WATER PIPING	Each	485.00
.3	DOMESTIC WATER SOFTENER	Each	910.00
.4	PLUMBING FIXTURES (Drain, Waste and Vent Not Included - See 1504.5)		
	Porcelain Enameled (unless otherwise noted) with accessories		
	Bathtub - 5' - Cast Iron - Enameled	Each	1,180.00
	Steel - Enameled	Each	600.00
	Fiberglass	Each	620.00
	5' x 5' - Cast Iron - Enameled	Each	1,540.00
	Bathtub/ Whirlpool - Fiberglass 6'	Each	4,200.00
	Bidet	Each	1,030.00
	Drinking Fountain - Enameled Wall Hung	Each	630.00
	SS Recessed	Each	760.00
	Garbage Disposal - 1/2 Horsepower	Each	175.00
	Lavatories - 19" x 17"- Oval	Each	315.00
	20" x 18" - Rectangular	Each	360.00
	27" x 23" - Oval Pedestal	Each	780.00
	18" - SS Bowl	Each	400.00
	Add for Handicap Lavatory	Each	196.00
	Shower & Tub - Combined - Fiberglass	Each	1,440.00
	Shower/Stall - Fiberglass - 32" x 32"	Each	852.00
	Fiberglass - 36" x 36"	Each	1,140.00
	Sink, Kitchen - 24" x 21" - Single	Each	550.00
	24" x 21" - SS Single	Each	385.00
	33" x 22" - Double	Each	610.00
	32" x 21" - SS Double	Each	460.00
	Sink, Laundry - Enameled	Each	610.00
	Plastic	Each	395.00
	Sink, Service - 24" x 21"	Each	745.00
	Urinals - Wall Hung	Each	665.00
	Stall or Trough 4'	Each	870.00
	Wash Fountain - Circular - Precast Terrazzo	Each	2,160.00
	Semi-circular - Precast Terrazzo	Each	2,160.00
	Water Closets - Floor Mounted - Tank Type	Each	480.00
	One Piece	Each	655.00
	Wall Mounted	Each	675.00
	Add for Colored Fixtures	Each	25%
	Add for Handicap Water Closet	Each	108.00
	See 1106 for Appliances		
.5	RESIDENTIAL PLUMBING		
	Average Domestic Fixture, Rough In and Trim	Each	540.00
	Average Cost of Drain, Waste, and Vent	Each	925.00
	Floor Drains - 4"	Each	490.00
.6	POOL EQUIPMENT - See 1318.0	-	-
.7	FOUNTAIN PIPING	-	-

		UNIT	COST
1505.0	**HEAT GENERATION**		
.1	FUEL HANDLING EQUIPMENT		-
.2	ASH REMOVAL SYSTEM - Electric	Each	9,100.00
	Hand	Each	5,940.00
.3	LINED BREECHINGS		-
.4	BOILERS		
	Package Type - 1,000 Lbs per Hour	Each	28,300.00
	Steam & Hot Water - 2,000 Lbs per Hour	Each	35,000.00
	5,000 Lbs per Hour	Each	58,700.00
	10,000 Lbs per Hour	Each	65,300.00
	15,000 Lbs per Hour	Each	90,200.00
	20,000 Lbs per Hour	Each	97,500.00
	25,000 Lbs per Hour	Each	134,000.00
	Cast Iron - 100 MBH	Each	2,570.00
	Hot Water or Steam - 200 MBH	Each	4,440.00
	300 MBH	Each	5,600.00
	600 MBH	Each	9,450.00
	1,000 MBH	Each	15,500.00
	Hot Water Heaters - 20 GPM	Each	660.00
	30 GPM	Each	850.00
	40 GPM	Each	950.00
	50 GPM	Each	1,280.00
1506.0	**REFRIGERATION**		
.1	REFRIGERANT COMPRESSORS		-
.2	CONDENSERS - Air Cooled - 10 Ton	Each	15,100.00
	20 Ton	Each	18,200.00
	30 Ton	Each	26,200.00
	40 Ton	Each	31,700.00
	60 Ton	Each	45,000.00
	80 Ton	Each	53,500.00
	Add for Water Cooled	Each	10%
.3	CHILLERS - Hermetic Centrifugal - 100 Ton	Each	57,800.00
	150 Ton	Each	59,950.00
	300 Ton	Each	85,200.00
	Add for Absorption Type	Each	15%
.4	COOLING TOWERS - 100 Ton	Each	8,200.00
	150 Ton	Each	15,100.00
	300 Ton	Each	40,800.00
1507.0	**HEAT TRANSFERS**		
.1	HOT WATER SPECIALTIES		-
.2	CONDENSATE PUMPS AND RECEIVER SETS		
	Outdoor - Air Cooled - 30,000 BTU	Each	2,060.00
	50,000 BTU	Each	2,780.00
	Indoor - 30,000 BTU	Each	1,050.00
	50,000 BTU	Each	1,450.00
.3	HEAT EXCHANGERS - 100 GPM	Each	1,240.00
	200 GPM	Each	1,330.00
	300 GPM	Each	2,900.00
	600 GPM	Each	6,300.00
	1,000 GPM	Each	7,880.00
.4	TERMINAL UNITS		
	Base Boards - Cast Iron Radiant - 7 1/4"	LnFt	24.50
	9 3/4"	LnFt	25.75
	Tube with Cover - 7 1/4" - 3/4"	LnFt	13.40
	7 1/4" - 1"	LnFt	14.50
	Tube with Cover - 9 3/4" - 3/4"	LnFt	16.70
	9 3/4" - 1"	LnFt	20.50
	14" Fin Tube - Steel	LnFt	25.10
	Copper	LnFt	30.50
	Convectors	Each	320.00
	Radiators	SqFt	8.90

1507.0	HEAT TRANSFERS, Cont'd...	UNIT	COST
.5	COILS	-	-
.6	UNIT HEATERS - 100,000 BTU	Each	750.00
	150,000 BTU	Each	1,030.00
	200,000 BTU	Each	1,180.00
.7	PACKAGED HEATING AND COOLING	-	2,400.00
.8	STEAM SPECIALTIES	-	540.00
1508.0	**AIR HANDLING AND DISTRIBUTION**		
.1	FURNACE - Duct - 100 MBH	Each	1,310.00
	200 MBH	Each	1,960.00
	300 MBH	Each	3,350.00
.2	FANS - 1,000 CFM	Each	145.00
	5,000 CFM	Each	610.00
	10,000 CFM	Each	1,020.00
.3	DUCT WORK - Galvanized - 26 Ga	Lb	3.35
	24 Ga	Lb	3.55
	22 Ga	Lb	4.05
	20 Ga	Lb	4.40
.4	DUCT ACCESSORIES		
	Constant Volume Duct	Each	600.00
	Reheat Constant Volume Duct	Each	735.00
	Diffusers & Registers - Sidewalk	SqFt	30.00
	Ceiling	SqFt	52.00
.5	RESIDENTIAL HEATING AND AIR CONDITIONING		
	Furnace	Each	1,460.00
	Central Air Conditioning	Each	690.00
	Humidifier	Each	340.00
	Air Cleaner	Each	550.00
	Thermostat	Each	144.00
	Attic Fan	Each	335.00
	Ventilating Damper	Each	123.00
	6" Flue	Each	377.00
.6	COMMERCIAL AIR CONDITIONING		
	Average	Ton	2,525.00
	High	Ton	3,900.00
	Low	Ton	1,240.00
1509.0	**SEWAGE DRAINAGE**		
.1	SEWAGE EJECTORS 100 GPM	Each	7,800.00
.2	GREASE INTERCEPTORS - 200 GPM	Each	9,100.00
	300 GPM	Each	17,100.00
.3	LIFT STATIONS - STEEL, CONCRETE OR FIBERGLASS		
	WITH GENERATOR - 100 GPM	Each	78,000.00
	200 GPM	Each	134,000.00
	500 GPM	Each	196,000.00
.4	SEPTIC TANKS AND DRAIN FIELD		
	Concrete Tanks with Excav. - 1,000 Gallons	Each	1,560.00
	2,000	Each	3,400.00
	5,000	Each	7,200.00
	10,000	Each	14,200.00
	Drain Fields	LnFt	5.00
.5	SEWAGE TREATMENT EQUIPMENT - STEEL OR CONCRETE		
	5,000 Gallons	GPD	10.00
	50,000	GPD	3.60
	100,000	GPD	3.40
.6	AERATION EQUIPMENT - STEEL - 1,000 Gallons	Each	18,500.00
	CAPACITY PER DAY 2,000	Each	27,800.00
	5,000	Each	37,500.00
	10,000	Each	45,200.00
.7	SLUDGE DIGESTION		-
1510.0	**CONTROLS AND INSTRUMENTATION** - No cost for items listed on Index		-
1511.0	**TESTING, ADJUSTING AND BALANCING**		

Costs of Electrical Work are priced as total contractors' costs with an overhead and fee of 20% included.

1601.0 BASIC MATERIALS AND METHODS

.1 CONDUITS (Incl Fittings/Supports)

COST

.11 Rigid

	UNIT	1/2"	3/4"	1-1/2"	2"	3"	4"	5"	6"
Galvanized - Rigid	LnFt	5.65	7.00	10.90	15.00	26.00	37.00	62.50	86.00
EMT	LnFt	3.70	4.75	6.45	8.20	14.50	20.60	-	-
PVC	LnFt	3.30	3.55	5.05	6.10	11.90	15.80	-	-
Add for Running Exposed									

.12 Flexible

	UNIT	1/2"	3/4"	1-1/2"	2"	3"	4"	5"	6"
Greenfield	LnFt	3.40	3.85	7.40	8.95	17.80	-	-	-
Sealtite	LnFt	5.00	6.20	13.00	16.70	36.00	44.00	-	-

.13 Ducts

	UNIT	1/2"	3/4"	1-1/2"	2"	3"	4"	5"	6"
Fibre	LnFt	-	-	-	-	4.40	4.75	5.40	6.00
Transite	LnFt	-	-	-	-	4.52	5.25	5.85	6.50

Cost for Conduits includes Labor, Material, Equipment and Fees and Installed in Wood or Steel Construction.

.2 BOXES, FITTINGS AND SUPPORTS (In Above)

COST

.3 WIRE AND CABLE

	UNIT	Copper T.H.W.
Wire - 3 - #14 - 600 Volt	LnFt	1.75
3 - #12 - 600 Volt	LnFt	2.15
3 - #10 - 600 Volt	LnFt	3.10
3 - # 8 - 600 Volt	LnFt	3.70
3 - # 4 - 600 Volt	LnFt	5.35
3 - # 2 - 600 Volt	LnFt	7.10
3 - 1/0	LnFt	9.75
3 - 2/0	LnFt	11.50
3 - 3/0	LnFt	12.80
3 - 4/0	LnFt	15.30
3 - 250 - MCM	LnFt	18.00
3 - 300 - MCM	LnFt	20.30
3 - 350 - MCM	LnFt	22.50
3 - 400 - MCM	LnFt	25.95
3 - 500 - MCM	LnFt	28.10

		ARMORED COPPER		NON-METALLIC	
	UNIT	2 Wire	3 Wire	2 Wire	3 Wire
Cable - #14 bx - 600 Volt	LnFt	3.50	3.90	1.65	3.00
#12 bx - 600 Volt	LnFt	3.70	4.20	1.80	3.10
#10 bx - 600 Volt	LnFt	4.30	5.15	2.30	3.40
# 8 bx - 600 Volt	LnFt	-	6.40	-	3.70
# 6 bx - 600 Volt	LnFt	-	7.20	-	4.15
# 4 bx - 600 Volt	LnFt	-	8.25	-	4.95
Add for Work over 10'					10%
Add for Work in Masonry Construction					10%
Add for Work in Concrete Construction					20%

.4 SWITCH GEAR

Metal Clad 5 KV	22,500.00
15 KV	27,200.00
Add for Outdoor Type	4,500.00

1601.0   BASIC MATERIALS AND METHODS, Cont'd...

		UNIT	COST
.5	SWITCHES AND RECEPTACLES		
	Switches		
	Standard Toggle - 1 Pole	Each	72.00
	4 W	Each	128.00
	4 WL	Each	145.00
	Safety - 30 Amp	Each	185.00
	60 Amp	Each	230.00
	100 Amp	Each	370.00
	200 Amp	Each	565.00
	Three-way -	Each	113.00
	Add for Dimmer - 600 Watt	Each	82.00
	Add for Dimmer - 1500 Watt	Each	175.00
	Receptacles		
	Single	Each	72.00
	Duplex - GFI	Each	92.00
	Add for Dedicated - Separate Circuit	Each	155.00
	Add for Dedicated - Ground	Each	20.50
	Add for Dedicated - Fourplex	Each	72.00
	Clock, Lamps, etc.	Each	74.00
	Add for Weatherproof	Each	18.50
	Plates - Stainless Steel	Each	10.20
	Brass	Each	15.50
.6	TRENCH DUCT AND UNDERFLOOR DUCT (Including Fittings)		
	Underfloor Duct - Average		
	Single Cell - 1 1/4" x 3 1/8"	LnFt	18.50
	1 1/2" x 6 1/2"	LnFt	23.20
	Double Cell - 1 1/4" x 3 1/8"	LnFt	26.70
	1 1/2" x 6 1/2"	LnFt	46.00
	Wireway - Cover Type 2 1/2" x 2 1/2" x 1"	Each	19.60
	2 1/2" x 2 1/2" x 2"	Each	26.75
	2 1/2" x 2 1/2" x 3"	Each	32.00
	4" x 4" x 1"	Each	24.75
	4" x 4" x 2"	Each	31.00
	4" x 4" x 3"	Each	39.00
.7	MOTOR WIRING (Including Controls)		
	10 H.P.	Each	4,000.00
	20 H.P.	Each	4,100.00
	30 H.P.	Each	4,600.00
	40 H.P.	Each	4,900.00
	50 H.P.	Each	5,800.00
	100 H.P.	Each	7,600.00
	200 H.P.	Each	11,600.00
.8	MOTOR STARTERS (600 Volt)		
	Magnetic - 10 H.P.	Each	490.00
	Non-Reversing - 20 H.P.	Each	580.00
	30 H.P.	Each	865.00
	40 H.P.	Each	1,020.00
	50 H.P.	Each	1,450.00
	100 H.P.	Each	2,250.00
	200 H.P.	Each	4,500.00
	Add for Reversing Type	Each	80%
	Add for Explosion Proof	Each	40%
	Add for Motor Circuit Protectors	Each	25%
	Add for Fused Switches	Each	10%
	Manual - 1 H.P.	Each	122.00
	3 H.P.	Each	175.00
	7 1/2 H.P.	Each	205.00

1601.0 BASIC MATERIALS AND METHODS, Cont'd...

		UNIT	COST
.9	CABLE TRAY (Aluminum or Galvanized)		
	Ladder or Punched - 6" wide	LnFt	28.00
	12" wide	LnFt	32.00
	18" wide	LnFt	34.00
	24" wide	LnFt	36.00

1602.0 POWER GENERATION

	UNIT	COST
Emergency Generators (3-Phase, 4-Wire) (120/208 Volt)		
15 k.w.	Each	13,400.00
30 k.w.	Each	26,600.00
50 k.w.	Each	33,000.00
75 k.w.	Each	36,000.00
100 k.w.	Each	46,300.00
300 k.w.	Each	77,000.00

1603.0 HIGH VOLTAGE DISTRIBUTION - Power Companies

1604.0 SERVICE AND DISTRIBUTION

		UNIT	COST
.1	METERING		
.2	GROUNDING (Including Rods and Accessories)		
	1/2" x 10'	Each	99.00
	5/8" x 10'	Each	108.00
	3/4" x 10'	Each	124.00
.3	SERVICE DISCONNECTS - 100 amp	Each	305.00
	200 amp	Each	565.00
	400 amp	Each	925.00
	600 amp	Each	1,800.00
.4	DISTRIBUTION SWITCHBOARDS (240 Volt) - 400 amp	Each	3,600.00
	800 amp	Each	3,700.00
	1600 amp	Each	5,350.00
.5	LIGHT AND POWER PANELS (3-Phase, 4-Wire) (120/208 Volt)		
	12 circuit	Each	740.00
	16 circuit	Each	885.00
	20 circuit	Each	1,030.00
	24 circuit	Each	1,230.00
	30 circuit	Each	1,440.00
	36 circuit	Each	1,600.00
	Add for Circuit Breakers	Each	235.00
.6	TRANSFORMERS		
	Current - 400 amp	Each	3,200.00
	600 amp	Each	3,500.00
	800 amp	Each	3,900.00
	1000 amp	Each	4,300.00

Light and Power	UNIT	SINGLE-PHASE 120/ 240 Volt	3-PHASE 120/ 208 Volt
Dry Type - 3 kva	Each	370.00	545.00
5 kva	Each	535.00	775.00
10 kva	Each	750.00	1,000.00
15 kva	Each	1,050.00	1,390.00
30 kva	Each		1,850.00
45 kva	Each		2,370.00
75 kva	Each		2,575.00
Oil Filled - 150 kva	Each	-	5,750.00
300 kva	Each	-	10,900.00
500 kva	Each	-	14,200.00

1605.0 LIGHTING

	UNIT	COST
.1 LIGHT FIXTURES		
Fluorescent		
Surface		
Industrial - 48" - 1 lamp R.S.	Each	108.00
H.O.	Each	135.00
2 lamp R.S.	Each	128.00
H.O.	Each	165.00
4 lamp R.S.	Each	144.00
H.O.	Each	245.00
96" - 1 lamp H.O.	Each	143.00
R.S.	Each	245.00
2 lamp H.O.	Each	122.00
H.O.	Each	255.00
Commercial - 12" x 48" - 2 lamp R.S.	Each	145.00
24" x 24" - 4 lamp R.S.	Each	190.00
24" x 48" - 4 lamp R.S.	Each	256.00
48" x 48" - 4 lamp R.S.	Each	380.00
12" x 48" - 2 lamp R.S.	Each	180.00
Recessed - 24" x 24" - 4 lamp R.S.	Each	195.00
24" x 48" - 4 lamp R.S.	Each	227.00
48" x 48" - 4 lamp R.S.	Each	395.00
Add for Electronic Ballasts	Each	23.00
Incandescent - 100 watt	Each	94.00
200 watt	Each	108.00
300 Watt	Each	125.00
Quartz - 500 watt	Each	170.00
Sodium - Low Pressure - 35 watt	Each	300.00
55 watt	Each	440.00
Explosion Proof		
Fluorescent- 48" - 2 lamp R.S.	Each	1,060.00
Incandescent - 48" - 200 watt	Each	340.00
Mercury Vapor - 48" - 200 watt	Each	670.00
Emergency - Nickel Cadmium Battery	Each	720.00
Lead Battery	Each	370.00
Exit - Standard	Each	155.00
Exit - Low Voltage - Battery	Each	190.00
Porcelain	Each	73.00
.2 LAMPS		
Fluorescent - 48" - 40 watt	Each	6.25
48" - 60 watt	Each	8.00
96" - 75 watt	Each	10.80
Incandescent - 25 to 100 watt	Each	2.25
400 watt	Each	3.60
Metal Halide - 175 watt	Each	67.00
400 watt	Each	72.00
1000 watt	Each	150.00
Sodium - 300 watt	Each	90.00
400 watt	Each	100.00
1000 watt	Each	226.00
Add for Disposal	Each	1.00

	UNIT	COST
1606.0 SPECIAL SYSTEMS		
LIGHTNING PROTECTION		
Ground Rod and Copper Point - 3/4" to 10"	Each	90.00
Ground Clamps - 2"	Each	110.00
Wire - #2	LnFt	1.15
1607.0 COMMUNICATION SYSTEMS		
.1 TELEPHONE EQUIPMENT		
Cabinets 24" x 36"	Each	320.00
36" x 48"	Each	485.00
Lines	Each	285.00
Phones	Each	162.00
Station	Each	600.00
.2 FIRE ALARM EQUIPMENT		
Control Panel 10 zones	Each	1,500.00
20 zones	Each	2,575.00
Annunciator Panel 10 zones	Each	575.00
Station	Each	245.00
Horn	Each	310.00
Heat Detector	Each	185.00
Smoke Detector	Each	205.00
.3 SECURITY SYSTEMS		
Standard Contact	Each	220.00
Audible Alarm	Each	270.00
Photo Electric Sensor - 500' Range	Each	455.00
Monitor Panel - Standard	Each	515.00
High Security	Each	690.00
Vibration Sensor	Each	195.00
Audio Sensor	Each	138.00
Motion Sensor	Each	330.00
.4 CLOCK PROGRAM EQUIPMENT (Control Panel and Clock with Buzzer)		
Master Time Clock 10 rm - average	Each	455.00
20 rm - average	Each	440.00
30 rm - average	Each	420.00
Time Clock and Cards	Each	1,180.00
Add for Electronic and Bells	Each	20%
.5 INTER-COMMUNICATION EQUIPMENT		
Individual Stations	Each	140.00
Board - Master	Each	570.00
.6 SOUND EQUIPMENT		
.61 Public Address System (Control and Speakers)		
3 to 6 Speakers - average each speaker	Each	545.00
7 to 9 Speakers - average each speaker	Each	495.00
10 to 20 Speakers - average each speaker	Each	430.00
Add for Tape / CD	Each	495.00
.62 Nurses & School Call systems (Panel & Stations)		
10 Stations or Rooms - average each room	Each	700.00
20 Stations or Rooms - average each room	Each	640.00
30 Stations or Rooms - average each room	Each	555.00
Add per Light	Each	67.00

1607.0	COMMUNICATIONS SYSTEMS, Cont'd...	UNIT	COST
.7	TELEVISION ANTENNA SYSTEMS		
	School Systems - 10 outlets	Each	335.00
	20 outlets	Each	330.00
	30 outlets	Each	305.00
	Apartment Houses - 20 outlets	Each	150.00
	30 outlets	Each	140.00
	50 outlets	Each	135.00
.8	CLOSED CIRCUIT T.V. STATIONS		
	Recorder	Each	3,700.00
	Camera	Each	1,550.00
	Surveillance	Each	1,960.00
.9	DOCTOR'S REGISTER	Each	6,800.00
	Add for Recording Register	Each	6,000.00

1608.0	HEATING AND COOLING		
.1	SNOW MELTING EQUIPMENT - 480 volt	SqFt	9.80
.2	ELECTRIC HEATING EQUIPMENT		
	Baseboard - 500 watt	LnFt	37.00
	750 watt	LnFt	36.50
	1000 watt	LnFt	35.00
	Cable Heating	SqFt	16.50
	Unit Heaters - 1500 watt	Each	215.00
	2500 watt	Each	310.00

1609.0	CONTROLS AND INSTRUMENTATION		
.1	LIGHTING CONTROL EQUIPMENT		-
.2	DIMMING EQUIPMENT		-
.3	TEMPERATURE CONTROL EQUIPMENT		-

1610.0	TESTING		-

1611.0	RESIDENTIAL WORK		
	100-Amp Service	Each	860.00
	200-Amp Service - Single Phase	Each	1,300.00
	Three Phase	Each	2,360.00
	Air Conditioner	Each	280.00
	Dimmer - 600 Watt	Each	56.00
	Dishwasher	Each	77.00
	Doorbell	Each	56.00
	Dryer - Electric	Each	108.00
	Dryer or Washer - Gas	Each	78.00
	Duplex Receptacle	Each	27.00
	Dedicated	Each	72.00
	Exhaust Fan - Bath	Each	93.00
	Kitchen	Each	98.00
	Fixture, Average	Each	93.00
	Porcelain	Each	89.00
	Furnace	Each	78.00
	Garbage Disposal	Each	75.00
	Paddle Fan	Each	202.00
	Phone - Data Box	Each	43.00
	Range Circuit	Each	135.00
	Smoke Detectors	Each	62.00
	Switches - Single	Each	27.00
	3-Way	Each	62.00
	Thermostat	Each	46.00
	Water Heater - 20 Amp 120 Volt	Each	118.00
	Whirlpool	Each	290.00

Average Square Foot Costs
2006

Housing (with Wood Frame)

Square Foot Costs do <u>not</u> include Garages, Land, Furnishings, Equipment, Landscaping, Financing, or Architect's Fee. Total Costs, except General Contractor's Fee, are included in each division. See Page 3 for Metropolitan Cost Variation Modifier. See Division 13 for Metal Buildings, Greenhouses, and Air-Supported Structures.

Division	Houses - 3 Bedrooms with Basement			Houses - 3 Bedrooms without Basement		
	Low Rent	Project	Custom	Low Rent	Project	Custom
1. General Conditions	$4.25	$4.35	$4.70	$5.50	$5.70	$6.10
2. Site Work	2.86	3.25	3.55	4.80	5.02	5.17
3. Concrete	3.75	3.95	4.16	6.73	6.95	7.63
4. Masonry	4.76	5.05	6.58	6.77	7.05	9.25
5. Steel	1.40	1.40	1.52	.70	.68	.70
6. Carpentry	15.50	15.24	17.08	22.60	22.50	24.35
7. Moisture & Therm. Prot.	3.76	3.75	3.95	5.25	5.42	5.92
8. Doors, Windows & Glass	7.50	7.76	8.16	9.90	10.20	10.67
9. Finishes	7.08	7.70	9.40	10.48	10.62	15.04
10. Specialties	.76	.76	.83	.83	.83	1.10
11. Equipment (Cabinets)	2.68	2.73	4.14	4.80	4.78	7.19
12. Furnishings	.55	.55	.60	.86	.86	.86
13. Special Construction	-	-	-	-	-	-
14. Conveying	-	-	-	-	-	-
15. Mechanical: Plumbing	5.70	5.68	6.70	8.35	8.34	10.32
Heating	3.72	3.85	4.00	5.18	5.50	7.78
Air Cond.	-	-	-	-	-	-
16. Electrical	3.25	3.25	3.93	4.85	4.85	5.75
SUB TOTAL	$67.52	$69.27	$79.30	$97.60	$99.30	$117.83
CONTRACTOR'S SERVICES	6.08	6.30	7.10	8.61	8.87	11.73
TOTAL AVERAGE Sq.Ft. Cost	**$73.60**	**$75.57**	**$86.40**	**$106.21**	**$108.17**	**$129.56**

	Unfinished Basement, Storage and Utility Areas included in Sq.Ft. Costs:			No Basement Utilities In Finished Area		
Average Cost/ Unit	$151,200	$153,240	$214,600	$125,800	$129,750	$193,500
Average Sq.Ft. Size/Unit	2,000	2,000	2,500	1,200	1,200	1,500
Add Architect's Fee	$ 6,725.00	$7,650.00	$17,160.00	$5,290.00	$6,820.00	$15,400.00

	COST	
	Sq. Ft.	Ea. Unit
Add for Finishing Basement Space For Bedrooms or Amusement Rooms, 12' x 20'	$28.00	$6,720
Add for Single Garage (no Interior Finish), 12' x 20'	$26.60	$6,240
Add for Double Garage (no Interior Finish), 20 x 20'	$28.00	$11,200
Add (or Deduct) for Bedrooms, 12' x 12'	$40.00	$5,760
Add per Bathroom with 2 Fixtures, 6' x 8'	$124.00	$5,950
Add per Bathroom with 3 Fixtures, 8' x 8'	$124.00	$7,900
Add for Fireplace	-	$6,500
Add for Central Air Conditioning	-	$5,200
Add for Drain Tiling	-	$2,800
Add for Well	-	$5,200
Add for Sewage System	-	$5,700

Average Square Foot Costs
2006

Housing

Square Foot Costs do <u>not</u> include Land, Furnishings, Equipment, Landscaping and Financing.

	Division	Apartments 2 & 3 Story (1)a	Apartments 2 & 3 Story (1)b	Apartments High Rise (2)	Motels & Hotels High Rise (3)	Motels & Hotels 1 & 2 Story (1)b	Motels & Hotels High Rise (2)
1.	General Conditions	$ 4.05	$ 4.30	$ 6.10	$ 6.50	$ 4.40	$ 5.60
2.	Site Work	2.55	2.50	3.60	3.55	5.24	4.05
3.	Concrete	2.10	2.10	20.05	13.52	2.48	20.55
4.	Masonry	1.65	10.26	2.70	2.58	10.30	13.24
5.	Steel	1.32	1.35	2.10	2.20	1.60	2.03
6.	Carpentry and Millwork	16.20	13.55	5.22	4.85	13.60	4.23
7.	Moisture Protection	3.48	3.70	1.90	1.88	3.38	1.88
8.	Doors, Windows and Glass	5.70	6.22	7.04	6.85	7.15	8.50
9.	Finishes	10.75	10.78	11.05	10.38	10.76	14.88
10.	Specialties	.83	.82	.88	.87	1.03	1.05
11.	Equipment	1.70	2.25	1.75	1.70	2.80	3.00
12.	Furnishings	.86	.92	.98	1.00	2.00	1.52
13.	Special Construction	-	-	-	-	-	-
14.	Conveying	-	2.37	3.62	3.80	-	4.20
15.	Mechanical:						
	Plumbing	6.68	6.68	6.55	6.40	7.62	10.20
	Heating and Ventilation	5.55	6.06	6.66	6.38	5.78	7.50
	Air Conditioning	2.00	2.00	2.10	2.05	2.50	4.88
	Sprinklers	-	-	2.12	1.90	1.96	2.02
16.	Electrical	7.03	7.04	8.00	8.42	9.40	9.90
	SUB TOTAL	$72.45	$82.90	$92.42	$85.35	$92.00	$119.23
	CONTRACTOR'S SERVICES	6.52	8.20	9.25	8.50	9.20	11.90
	Construction Sq.Ft. Cost	$78.97	$91.10	$101.67	$93.85	$101.20	$131.13
	Add for Architect's Fee	5.52	7.00	8.00	7.80	7.80	10.90
	TOTAL Sq.Ft. Cost	**$84.49**	**$98.10**	**$109.67**	**$101.65**	**$109.00**	**$142.03**
	Average Cost per Unit without Architect's Fee	$75,020	$86,545	$91.580	$75,080	$65,780	$85.085
	Average Cost per Unit with Architect's Fee	$80,275	$93,195	$98,775	$81,320	$70,850	$91,650
	Average Sq.Ft. Size per Unit	950	950	900	800	650	650
	Average Sq.Ft. Living Space	750	750	750	600	450	450

Add for Porches – Wood - $11.50 SqFt
Add for Interior Garages - $10.50 SqFt

(1)a Wall Bearing Wood Frame & Wood or Aluminum Facade
(1)b Wall Bearing Masonry & Wood Joists & Brick Facade
(2) Concrete Frame
(3) Post Tensioned Concrete Slabs, Sheer Walls, Window Wall Exteriors, and Drywall Partitions.

Average Square Foot Costs
2006

Commercial

Square Foot Costs do not include Land, Furnishings, and Financing.

	Division	Remodeling Office Interior	Office 1-Story (1)	Office 2-Story (2)	Office Up To 5-Story (3)	High Rise (City) Above 5-Story (4)	Parking Ramp (3)
1.	General Conditions	$ 2.57	$ 4.20	$ 4.19	6.18	6.50	4.58
2.	Site Work & Demolition	1.50	2.10	2.10	2.10	1.96	1.40
3.	Concrete	-	6.45	10.03	11.05	7.50	23.79
4.	Masonry	-	10.47	9.90	14.96	3.70	1.50
5.	Steel	-	3.10	10.25	3.59	2.52	1.95
6.	Carpentry and Millwork	7.95	3.68	3.43	5.35	6.02	1.23
7.	Moisture Protection	-	6.40	3.20	2.14	2.69	2.98
8.	Doors, Windows and Glass	1.68	6.52	6.51	13.36	25.08	1.35
9.	Finishes & Sheet Rock	11.30	12.05	11.78	13.85	15.02	1.82
10.	Specialties	-	1.66	1.66	1.72	1.72	-
11.	Equipment	-	5.25	4.65	3.86	5.22	-
12.	Furnishings	-	1.10	1.30	1.43	1.55	-
13.	Special Construction	-	-	-	-	-	-
14.	Conveying	-	-	3.00	4.92	5.66	2.88
15.	Mechanical:						
	Plumbing	3.50	5.29	5.67	6.56	6.82	2.30
	Heating and Ventilation	1.73	6.45	6.29	8.15	8.20	-
	Air Conditioning	1.67	6.40	6.60	7.80	7.80	-
	Sprinklers	-	1.98	1.98	2.03	2.03	-
16.	Electrical	6.75	8.31	9.15	10.06	11.89	2.62
	SUB TOTAL	$38.65	$91.41	$101.69	$119.11	$ 121.86	$48.40
	CONTRACTOR'S SERVICES	3.86	8.22	9.63	10.71	10.95	4.82
	Construction Sq.Ft. Cost	$42.51	$99.63	$111.32	$129.82	$132.81	$53.25
	Add for Architect's Fee	4.25	69.87	7.80	9.00	10.60	4.24
	TOTAL Sq.Ft. Costs	$46.76	$169.50	$119.12	$138.82	$143.41	$57.49
	Deduct for Tenant Areas Unfinished	-	-	22.00	25.00	29.00	-
	Per Car Average	-	-	-	-	-	$16,000
	Add: Masonry Enclosed Ramp	-	-	-	-	-	6.00
	Add: Heated Ramp	-	-	-	-	-	5.80
	Add: Sprinklers for Closed Ramps	-	-	-	-	-	2.30
	Deduct for Precast Concrete Ramp	-	-	-	-	-	8.40

(1) Wall Bearing Masonry and Steel Joists and Decks – No Basement
(2) Wall Bearing Masonry and Precast Concrete Decks – No Basement
(3) Concrete Frame
(4) Steel Frame and Window Wall Facade

Average Square Foot Costs
2006

Commercial and Industrial

Square Foot Costs do not include Land, Furnishings, Landscaping, and Financing.
See Division 13 for Metal Buildings

	Division	Store 1-Story (1)	Store 2-Story (2)	Production 1-Story (1)	Warehouse 1-Story (1)	Warehouse and 1-Story Office (1)	Warehouse and 2-Story Office (3)
1.	General Conditions	$ 3.80	$ 3.82	$ 3.70	$ 3.30	$ 3.40	$ 3.10
2.	Site Work & Demolition	1.90	1.48	1.72	1.67	2.00	1.60
3.	Concrete	5.66	10.30	5.75	5.56	6.20	9.45
4.	Masonry	5.20	8.85	6.90	6.92	7.65	4.20
5.	Steel	9.82	5.15	9.82	9.60	10.42	7.87
6.	Carpentry	2.78	2.84	1.80	1.82	2.45	2.15
7.	Moisture Protection	4.50	2.61	4.95	4.75	4.78	2.38
8.	Doors, Windows and Glass	5.32	5.00	4.15	2.22	4.50	3.95
9.	Finishes	6.10	5.75	5.27	1.87	3.28	5.05
10.	Specialties	1.58	1.60	1.20	1.15	1.15	1.15
11.	Equipment	1.48	1.70	1.60	-	.90	1.03
12.	Furnishings	-	-	-	-	-	-
13.	Special Construction	-	-	-	-	-	-
14.	Conveying	-	3.52	-	-	-	-
15.	Mechanical:						
	Plumbing	4.15	3.50	4.36	2.51	3.17	2.95
	Heating and Ventilation	4.18	4.15	3.22	3.33	4.48	5.07
	Air Conditioning	4.30	4.64	3.06	-	1.15	1.05
	Sprinklers	1.96	1.90	1.85	1.90	2.02	1.96
16.	Electrical	7.70	7.35	8.15	4.90	5.05	4.62
	SUB TOTAL	$70.43	$74.16	$67.50	$51.50	$62.60	$57.58
	CONTRACTOR'S SERVICES	7.05	7.45	6.75	5.15	6.26	5.76
	Construction Sq.Ft. Cost	$77.48	$81.61	$74.25	$56.65	$68.86	$63.34
	Add for Architect's Fees	5.50	5.77	5.20	4.00	4.90	4.44
	TOTAL Sq.Ft. Cost	$82.98	$87.38	$79.45	$60.65	$73.76	$67.78

(1) Wall Bearing Masonry and Steel Joists and Decks – Brick Fronts
(2) Wall Bearing Masonry and Precast Concrete Decks – Brick Fronts
(3) Precast Concrete Wall Panels, Steel Joists, and Deck

Average Square Foot Costs
2006

Educational and Religious

Square Foot Costs do not include Land, Furnishings, Landscaping and Financing.

	Division	Elementary School 1-Story (1)	High School 2-Story (1)	Vocational School 3-Story (1)	College Multi-Story (2)	Gym (3)	Dorm 3-Story (1)	Church (3)
1.	General Conditions	$ 6.65	$ 6.35	$ 6.45	$ 6.30	$ 5.20	$ 5.75	6.20
2.	Site Work	3.44	3.60	3.50	3.40	1.72	2.25	3.31
3.	Concrete	5.75	6.00	4.33	20.50	6.05	4.55	5.50
4.	Masonry	12.95	11.70	12.70	13.33	13.12	11.87	26.65
5.	Steel	9.90	11.20	11.76	4.12	12.12	10.50	1.78
6.	Carpentry	3.96	3.80	3.52	2.50	2.20	3.41	7.20
7.	Moisture Protection	5.15	3.40	3.95	3.97	5.44	4.75	5.65
8.	Doors-Windows-Glass	6.30	6.30	6.65	8.32	5.82	5.42	7.62
9.	Finishes	10.80	8.85	9.90	8.75	7.50	10.33	10.75
10.	Specialties	1.65	1.60	1.60	1.98	1.60	1.55	1.20
11.	Equipment	4.45	4.40	6.60	5.75	8.65	5.46	7.85
12.	Furnishings	-	.93	1.14	1.21	-	1.15	4.56
13.	Special Construction	-	-	-	-	-	-	-
14.	Conveying	-	-	3.27	3.27	-	3.25	-
15.	Mechanical:							
	Plumbing	8.20	8.38	8.38	11.28	5.20	7.70	5.30
	Heating-Ventilation	11.80	11.45	12.95	11.85	7.48	8.00	8.22
	Air Conditioning	-	-	-	-	-	-	6.23
	Sprinklers	2.00	2.00	2.00	2.02	1.02	2.01	
16.	Electrical	12.50	12.86	13.40	13.60	8.50	9.72	8.50
	SUB TOTAL	$105.50	$102.82	$112.10	$122.15	$91.62	$97.67	116.52
	CONTRACTOR'S SERVICES	10.55	10.25	11.20	12.42	9.26	9.76	11.60
	Constr. Sq.Ft. Cost	$116.05	$113.07	$123.30	$134.57	$100.88	$107.43	128.12
	Add Architect's Fees	8.12	7.89	8.62	9.36	7.09	7.52	9.00
	TOTAL Sq.Ft. Cost	$124.17	$120.96	$131.92	$143.93	$107.97	$114.95	$137.12

(1) Wall Bearing Masonry and Steel Joists and Decks
(2) Concrete Frame and Masonry Facade
(3) Masonry Walls and Wood Roof

Average Square Foot Costs
2006

Medical and Institutional

Square Foot Costs do not include Land, Landscaping, Furnishings, and Financing.

		Clinic 1-Story (1)	Doctors Office 1-Story (1)	Hospital 1-Story (1)	Hospital Multi-Story (2)	Nursing Home 1-Story (1)	Housing for Elderly 1-Story (1)
1.	General Conditions	$ 6.80	$ 6.90	$ 9.15	$ 9.25	$ 8.30	$ 7.60
2.	Site Work	2.15	2.10	3.31	3.35	3.50	3.65
3.	Concrete	7.00	6.95	7.12	21.74	7.40	7.50
4.	Masonry	11.95	10.88	19.65	13.80	14.05	10.34
5.	Steel	8.60	8.50	10.41	2.31	2.32	1.97
6.	Carpentry	6.80	6.72	5.92	3.85	4.13	3.93
7.	Moisture Protection	4.94	4.88	5.26	2.22	5.25	5.22
8.	Doors, Windows and Glass	7.96	8.40	10.50	7.03	9.15	7.30
9.	Finishes	10.70	10.45	13.55	13.12	13.03	10.95
10.	Specialties	1.48	1.48	1.60	1.52	1.52	1.49
11.	Equipment	8.00	7.52	1.92	12.10	6.07	2.70
12.	Furnishings	1.20	1.20	1.27	1.35	1.13	2.40
13.	Special Construction	.88	.87	1.75	1.75	.87	.88
14.	Conveying	-	-	-	6.52	-	6.12
15.	Mechanical:						
	Plumbing	10.92	11.73	17.73	17.25	11.82	11.48
	Heating and Ventilation	9.60	8.55	12.55	12.93	6.50	5.25
	Air Conditioning	7.40	7.21	12.90	11.20	7.48	2.83
	Sprinklers	2.02	1.98	2.00	2.02	2.28	2.00
16.	Electrical	14.35	12.07	21.54	20.26	13.10	11.52
	SUB TOTAL	$122.75	$118.39	$158.13	$163.57	$117.90	$105.13
	CONTRACTOR'S SERVICES	13.00	11.85	15.82	16.12	11.78	10.52
	Construction Sq.Ft. Cost	$135.75	$130.24	$173.95	$179.69	$129.68	$115.65
	Add for Architect's Fees	10.00	9.13	12.16	12.45	9.07	8.05
	TOTAL Sq.Ft. Cost	$145.75	$139.37	$186.11	$192.14	$138.75	$123.70

Average Cost to Remodel Approximately 50% Cost of New

---

(1) Wall Bearing Masonry and Steel Joists and Decks
(2) Concrete Frame and Masonry Façade

APPENDIX

Average Square Foot Costs
2006

Government and Finance

Square Foot Costs do <u>not</u> include Land, Landscaping, Furnishings, and Financing.

		Bank & Drive-Thru 1-Story (1)	Bank & Insur. Bldg High-Rise (3)	Govt. Office Bldg Multi-Story (2)	Post Office 1-Story (1)	Court House 3-Story (2)	Jail and Prison (2)
1.	General Conditions	$ 7.60	$ 8.70	$ 8.70	$ 7.07	$ 8.55	$ 10.53
2.	Site Work	4.10	3.73	3.15	2.61	2.85	3.78
3.	Concrete	9.75	23.75	25.10	10.67	21.32	29.34
4.	Masonry	14.42	10.05	16.75	10.03	16.70	18.30
5.	Steel	4.83	2.17	2.63	3.80	2.27	5.45
6.	Carpentry	4.62	4.18	4.29	3.78	2.68	4.74
7.	Moisture Protection	3.68	4.21	4.40	3.73	4.64	5.33
8.	Doors, Windows and Glass	10.35	24.12	10.66	8.15	7.65	9.92
9.	Finishes	14.88	16.50	17.60	15.02	14.22	12.10
10.	Specialties	1.60	1.60	3.08	1.65	2.70	6.50
11.	Equipment	12.72	8.46	8.46	11.21	10.65	12.42
12.	Furnishings	1.43	1.31	-	1.93	-	-
13.	Special Construction	-	-	-	-	-	-
14.	Conveying	-	2.67	3.45	1.76	2.52	3.38
15.	Mechanical						
	Plumbing	6.68	7.82	7.74	8.38	10.63	18.45
	Heating and Ventilation	5.91	7.36	7.90	5.71	8.45	15.02
	Air Conditioning	4.75	5.40	6.88	6.85	12.02	12.48
	Sprinklers	2.15	2.04	1.99	2.05	2.05	2.04
16.	Electrical	13.38	16.10	16.70	12.72	15.92	25.42
	SUB TOTAL	$122.85	$150.17	$149.48	$117.12	$145.82	$195.20
	CONTRACTOR'S SERVICES	11.07	13.95	13.05	10.53	13.16	17.55
	Construction Sq.Ft. Cost	$133.92	$164.12	$162.53	$127.65	$158.98	$212.75
	Add for Architect's Fees	9.38	11.82	11.06	8.94	11.20	17.04
	TOTAL Sq.Ft. Cost	$143.30	$175.94	$173.59	$136.59	$170.18	$229.79

---

(1) Wall Bearing Masonry and Precast Concrete Decks
(2) Concrete Frame and Masonry Facade
(3) Concrete Frame and Window Wall

U.S. Measurements & Metric Equivalents – (For Construction)

Weights & Measures (U.S. System)

Length	1 Mile = 320 Rods = 1,760 Yards = 5,280 Feet
Area	1 Mile = 640 Acres = 102,400 Sq. Rods = 3,097,600 Sq. Yards
Area	1 Acre = 160 Sq. Rods = 4,840 Sq. Yards = 43,560 Sq. Feet
Volume	1 Cu. Yard = 27 Cu. Feet = .211 Cord
Capacity	1 Gallon = 4 Quarts = 8 Pints = .13337 Cu. Feet
Weight & Mass	1 Ton = 2,000 Pounds = 32,000 Ounces

U.S. to Metric – Area, Length, Volume & Mass Conversion Factors

Quantity	From U.S.	To Metric	Multiply By
Length	mile	km	1.609 344
	yard	m	0.914 4
	foot	m	0.304 8
	inch	mm	25.4
Area	square mile	km^2	2.590 00
	acre	m^2	4 046.87
	square yard	m^2	0.836 127
	square foot	m^2	0.092 903
	square inch	mm^2	645.16
Volume	cubic yard	m^3	0.764 555
	cubic foot	m^3	0.028 316 85
Mass-(Weight)	lb	kg	0.453
Mass-(Length)	plf	kg/m	1.488
Mass-(Square)	psf	kg/m^2	4.882
Mass-(Cube)	pcf	kg/m^3	16.018

Length Conversion Tables from Fractions to Decimals to Millimeters

Fraction	Decimal	Millimeters	Fraction	Decimal	Millimeters
1/16"	.0625	1.588	9/16"	.5625	14.288
1/8"	.125	3.175	5/8"	.625	15.875
3/16"	.1875	4.763	11/16"	.6874	17.463
1/4"	.250	6.350	¾"	.750	19.050
5/16"	.3125	7.938	13/16"	.8125	20.638
3/8"	.375	9.525	7/8"	.875	22.225
7/16"	.437	11.113	15/16"	.9375	23.813
1/2"	.500	12.700	1"	1.000	25.400

Weights – Construction Products

Substance	Wt.Lb. per Cu.Ft.	Substance	Wt.Lb. per Cu.Ft.	Substance	Wt.Lb. per Cu.Ft.
Aluminum	165	Lead	710	Concrete	144
Brass	534	Paper	58	Cement	90
Bronze	509	Glass	156	Clay	100
Copper	556	Granite	165	Sand	100
Iron	450	Limestone	160	Asphalt	81
Steel	490	Brick	140	Wood	40

Computation of Area

Square or Rectangular	=	Length x Width or Height
Triangle	=	Base x ½ Height
Circumference of Circle	=	Diameter x 3.1416
Area of Circle	=	Radius Squared x 3.1416
Cube of a Column or Cylinder	=	Radius Squared x 3.1416 x Depth
Cube of a Solid	=	Length x Width x Depth
Area of Ellipse	=	½ Length x ½ Height x 3.1416

UNITED STATES INFLATION RATES
July 1, 1970 to June 30, 2005
AC&E PUBLISHING CO. INDEX

Start July 1

YEAR	AVERAGE %	LABOR %	MATERIAL %	CUMULATIVE %
1970-71	8	8	5	0
1971-72	11	15	9	11
1972-73	7	5	10	18
1973-74	13	9	17	31
1974-75	13	10	16	44
1975-76	8	7	15	52
1976-77	10	5	15	62
1977-78	10	5	13	72
1978-79	7	6	8	79
1979-81	10	5	13	89
1980-81	10	11	9	99
1981-82	10	11	10	109
1982-83	5	9	3	114
1983-84	4	5	3	118
1984-85	3	3	3	121
1985-86	3	2	3	124
1986-87	3	3	3	127
1987-88	3	3	4	130
1988-89	4	3	4	134
1989-90	3	3	3	137
1990-91	3	3	4	140
1991-92	4	4	4	144
1992-93	4	3	5	148
1993-94	4	3	5	152
1994-95	4	3	6	156
1995-96	4	3	6	160
1996-97	3	3	3	163
1997-98	4	3	4	167
1998-99	3	3	2	170
1999-2000	3	2	3	173
2000-2001	4	4	4	177
2001-2002	3	4	3	180
2002-2003	4	5	3	184
2003-2004	3	5	2	187
2004-2005	8	4	10	195
2005-2006	8	6	9	203